Advances in Sustainability Science and Technology

The book series aims at bringing together valuable and novel scientific contributions that address the critical issues of renewable energy, sustainable building, sustainable manufacturing, and other sustainability science and technology topics that have an impact in this diverse and fast-changing research community in academia and industry.

The areas to be covered are

- Climate change and mitigation, atmospheric carbon reduction, global warming
- Sustainability science, sustainability technologies
- Sustainable building technologies
- Intelligent buildings
- Sustainable energy generation
- Combined heat and power and district heating systems
- Control and optimization of renewable energy systems
- Smart grids and micro grids, local energy markets
- Smart cities, smart buildings, smart districts, smart countryside
- Energy and environmental assessment in buildings and cities
- Sustainable design, innovation and services
- Sustainable manufacturing processes and technology
- Sustainable manufacturing systems and enterprises
- Decision support for sustainability
- Micro/nanomachining, microelectromechanical machines (MEMS)
- Sustainable transport, smart vehicles and smart roads
- Information technology and artificial intelligence applied to sustainability
- Big data and data analytics applied to sustainability
- Sustainable food production, sustainable horticulture and agriculture
- Sustainability of air, water and other natural resources
- Sustainability policy, shaping the future, the triple bottom line, the circular economy

High quality content is an essential feature for all book proposals accepted for the series. It is expected that editors of all accepted volumes will ensure that contributions are subjected to an appropriate level of reviewing process and adhere to KES quality principles.

The series will include monographs, edited volumes, and selected proceedings.

Neha Sharma · Prithwis Kumar De

Towards Net-Zero Targets

Usage of Data Science for Long-Term Sustainability Pathways

Springer

Neha Sharma
Tata Consultancy Services
Pune, India

Prithwis Kumar De
Tata Consultancy Services
London, UK

ISSN 2662-6829 ISSN 2662-6837 (electronic)
Advances in Sustainability Science and Technology
ISBN 978-981-19-5243-2 ISBN 978-981-19-5244-9 (eBook)
https://doi.org/10.1007/978-981-19-5244-9

This Springer imprint is published by the registered company Springer Nature Singapore Pte Ltd.
The registered company address is: 152 Beach Road, #21-01/04 Gateway East, Singapore 189721, Singapore

Dedicated to our Families

Foreword by Padmashri Prof. Amitav Mallik

One of the critical challenges that societies, economies, industries, governments, and regulators face today is climate change due to the fast-growing global warming. Some of the prominent effects of climate change are extreme weather conditions such as flash floods, frequent cyclones, and droughts that are already happening and some long-term effects like glacial melting and sea-level rise affecting coastal cities. All these adverse effects not only lead to the loss of physical and financial assets but also play havoc on human lives and well-being. Therefore, it is imperative to make concerted efforts to prevent and mitigate the negative effect of climate change.

This book is an excellent effort to influence possible action through the use of Data Science which can be the basis of the collaboration among countries, multinational industries, civic-society organizations, governments, academia, and citizens. This will be imperative for preserving the earth's ecosystems for generations to come and ensure a better future for them.

The book has 10 chapters highlighting five domains and has made use of Open Data to find some insights. This book summarizes the climate change problem from the perspective of four significant sectors, namely, financial, non-renewable energy, domestic, and transport sectors. It provides solutions by leveraging open data and applying data science as well as machine learning techniques. The analysis of the financial sector shows that climate change due to CO_2 and methane emissions, etc., will keep increasing. The vast investments by banks and financial institutions in Coal Power, Coal Mining, Oil extraction, and even Fracked Oil & Gas have led to an unprecedented situation where the survival of humanity is getting threatened. An attempt has been made to predict average land temperature due to CO_2 emissions from fossil-fuel resources like coal, oil, natural gas, etc., by using time series data. The findings suggest that temperatures will keep rising in the coming years.

The authors have used unsupervised machine learning for the domestic sector to classify the Indian states and union territories into three clusters based on household emissions. The authors have also carried out the profiling of the clusters by considering population, per capita emissions, and per-square km emissions, which generates very fascinating insights. After extensive data analysis, it is concluded that as the fuel consumption on the highway and in cities increases or the vehicle size increases, the CO_2 emissions will also increase. On the other hand, the better the economy of a vehicle, the lower the CO_2 emissions. At the end, the authors have proposed a real-time carbon emissions calculator for personal computers, which would help to find out the carbon emissions dynamically.

Therefore, with great pleasure and satisfaction, I introduce this book on "Net Zero" to the community for whom it is most relevant. I would encourage the researchers, practitioners, and policymakers to contribute towards achieving "Net Zero" and build further on the impressive work that has been done.

Padmashri Prof. Amitav Mallik
Founding Trustee
Pune International Centre

Head
Energy Environment and Climate Change (EECC)
and Founder, Climate Collective Pune
Pune, India

Foreword by Dinanath Kholkar

The current climate crisis is clearly a wake-up call and we all need to be seriously concerned about our actions that harm the environment. Climate literacy is the need of the hour and has to influence consumer behavior. The climate change ecosystem and the role of data science are of prime importance to individuals, communities, and businesses. Growth and development need to be balanced with a concern for sustainability, and open data, machine learning, and relevant digital technologies will play a significant role in this journey.

A strong collaboration is required among industry, academia, policymakers, regulators, and citizens to combat climate change and make a positive contribution towards achieving the net-zero targets. Reading through the book, you will find some of the specific challenges as well as solutions regarding climate change in the financial, non-renewable energy, household, and transport sectors, through the lens of data science. I encourage the readers to make the most of the research carried out by the authors towards increasing climate literacy and mitigating climate risk.

Dinanath Kholkar
Senior Vice President and Global Head
of Partner Ecosystem and Alliances
Tata Consultancy Services Ltd.
Pune, India

Preface

The recent years have witnessed many unprecedented natural disasters due to significant changes in the earth's climate. Large sectors such as manufacturing, transportation, energy and utilities, consumer products and retail, etc., have a sizeable climate footprint. Their operations add to Green House Gas (GHG) concentration, which seems to be one of the reasons behind global warming and climate change. There have been many discussions around sustainability and climate change, and to find out ways to mitigate the negative impact. However, using technology levers to contain the adverse effects of climate change is rarely seen as the most significant catalyst. Recent development in Artificial Intelligence (AI) and Machine Learning (ML) coupled with improved open data can offer an exciting new front in the prevention of climate change. It can significantly benefit organizations and nations in reducing GHGs, improving energy efficiency, and reducing waste. Most of the available research in this area is qualitative in nature, where technology and data have not yet been leveraged.

In this book, the authors have made an attempt to shift the attention from merely identifying and analyzing climate risk to actively addressing it by leveraging technology. AI/ML is an enabling technology that has a critical role in reducing the carbon footprint. Chapter 1 outlines a brief description of various domains covered in this book and the carbon emitted by each one of them while performing their normal activities. The chapter highlights the usage of AI/ML to predict the climate change consequences arising due to investment in fossil fuel sectors, predicts CO_2 emissions from the transport sector, forecasts average land temperature due to non-renewable sources of energy, and segments Indian states on the basis of household carbon emissions. The remaining chapters dive deep into the individual domains and present insights which can guide us to arrest the negative impacts of climate change. The financial institutions, energy and transportation sectors, as well as individuals, can be encouraged to collect data from operations activities and every part of the value chain, and then use AI-powered data engineering to track emissions throughout their carbon footprint.

Financial Sector—Chap. 2 emphasizes the crucial role of the bank in the economic and social sphere as a provider of capital and decision-maker for the sectors to be financed. Nevertheless, they are also exposed to physical and transition risks due to climate change. The financial institutions may not be the major polluters themselves, but due to their financial intermediary role, they might have a relationship with companies or investment portfolios that are carbon-intensive in nature. Financial institutions can contribute towards sustainable development by carrying out responsible investments in companies that aim at reducing energy consumption and carbon emissions. After an exhaustive literature review and preliminary data analysis, it is observed that there is a direct relation between the investment by banks in various fossil fuel sectors and the consequences of climate change.

In Chap. 3, ML modeling has been carried out to predict the consequences related to climate change due to the investment by banks in fossil fuel sectors. The analysis shows that the climate change consequences and the investments by banks in different fossil fuel sectors are correlated. It has been concluded that climate change consequences like methane emissions, carbon emissions, etc., are adversely impacted due to the investment by banks in Tar Sands, Coal Power, Coal Mining, LNG, Fracked Oil and Gas, Offshore Oil and Gas, and Arctic Oil and Gas. We would recommend that the banks should phase out their investment in fossil fuel sectors in a strategic manner. There must be more disclosure of financial information by the banks, investors, and companies targeting carbon-intensive sectors.

Non-renewal Energy Sector—Chap. 4 highlights how the energy sector has been the most significant contributor to GHG emissions. It is a difficult and very challenging goal that several countries need to work on by balancing between advancement, progress, and development, as well as the climate change crisis. The aim of the research is to study India's energy sector and different sources of non-renewable energy, and to clearly distinguish their potential and contributions towards climate change. The interrelation between energy and climate change has been explored in this chapter. Increasing emissions have resulted in higher average land temperatures due to the extensive use of non-renewable energy sources for many decades.

Chapter 5 makes an attempt to predict trends in average land temperature due to CO_2 emissions from all the non-renewable energy resources like "coal", "oil", "natural gas", and "flaring". The chapter illustrates the step-by-step process of time series analysis that is used to forecast temperature values. Predictions with the help of a ML model reveal that if no significant and immediate steps are implemented, emissions from these sources of energy will continue to rise in the near future at the current rate of consumption and production. The findings indicate that the average land temperature in India is predicted to rise by 0.2 °C in the coming years.

Domestic Sector—GHG emissions from the domestic/household/residential sector in India are primarily contributed by electricity usage in households and fuel consumption such as LPG, kerosene, firewood, crop residues, and animal dung in urban and rural homes. Chapter 6 starts with an introduction to climate change, discusses global initiatives taken to tackle it, and highlights the emissions from the household sector in India. The chapter achieved the objective of comparing residential and primary energy household emissions to find out the similarity between the two.

Chapter 7 is focused to determine the direction of primary energy household emissions in India from LPG, Kerosene, and Diesel. Besides, it also classifies 35 Indian states and union territories by applying unsupervised machine learning algorithms based on their varying primary energy household emissions from CO_2, CH_4, and N_2O. This created three clusters, "Excessive Emitters", "High Emitters", and "Low Emitters", consisting of 1, 11, and 23 states, respectively. In addition, Kerosene and LPG have been identified as important factors in the formation of the cluster. Further, the profiling of the clusters has been carried out by considering population, per capita emissions, and per-square kilometer emissions. Uttar Pradesh is a part of the excessive emission cluster, but it has relatively low per capita and per-square kilometer emissions. Most of the union territories are members of the low emission cluster but have higher per-square kilometer emissions. Though Delhi belongs to the low emission cluster, it is the highest both in per capita and per-square kilometer emissions.

Transportation Sector—The transport sector is one of the highest GHG emitters and is responsible for the majority of air pollutant precursor emissions. A significant reason for these high percentages of GHGs emitted by the transport sector is the extensive consumption of fossil fuel which is used to power combustion engines. Chapter 8 is a detailed study of the available literature from understanding the impact of the transport sector as a significant contributor to climate change to the application of data science for combating the same. The study reveals that countries like the USA, the UK, Canada, and France are the front footers to solve the climate change problem caused by the transport sector.

Chapter 9 focuses on the analysis of data from various perspectives, presents the correlation between the attributes, and selects the features for building up a model. This chapter deliberates on building up various sophisticated ML models, evaluating their performances, and suggesting the best model for predicting CO_2 emissions. After extensive data analysis, it is concluded that, as the fuel consumption on the highway and in the city increases or the vehicle size increases, the CO_2 emissions also increase. On the other hand, the better the economy of a vehicle, the lower the CO_2 emissions.

Adverse Effects of AI—Chaps. 1–9 present how AI can be applied to combat climate risks, whereas Chap. 10 focuses on elucidating the negative aspects of AI towards climate change. To power AI systems, energy is required, which comes from electricity. Most of the electricity comes from non-renewable sources of energy, which could lead to more carbon dissipation. Inference on a ML emissions calculator has been made, and different combinations of hardware, region of compute, and cloud providers for machine learning models, have been analyzed to get a glimpse of their carbon footprint. Furthermore, this chapter proposes a real-time carbon emissions calculator for personal computers, which would help to find out the system's exploitation in terms of its carbon emissions dynamically.

In this book, the technology, AI/ML, open data, etc., have been extensively applied to propose cutting-edge models to achieve the net-zero carbon emissions targets set by the Paris Agreement on climate change. We hope this book helps to improve education, raise awareness, and build knowledge on climate change prevention and mitigation. We think researchers, policymakers, students, teachers, educational institutions, governments, companies, international organizations, etc., will benefit immensely by referring to this book. Moreover, we endeavor to provide a decarbonized environment and a better tomorrow to the next generation through this book.

Pune, India Neha Sharma
London, UK Prithwis Kumar De

Acknowledgements

Writing a book is more strenuous and demanding than we thought and more rewarding than we could have ever imagined. We, the authors of the book, take this opportunity to express our heartfelt gratitude to all those who have contributed to this book and supported us in numerous ways. We shall remain indebted to all of them. Though the book is authored by us, many people have helped to bring it to its present shape. We are particularly grateful to Mr. Aninda Bose, Senior Publishing Editor from Springer Nature, for his encouragement and continuous interest in our book. We extend our sincere thanks to Gokul Hiralal Bisen and Khaleel Ur Rehman Khan from Visvesvaraya National Institute of Technology, Nagpur, Maharashtra; Aryam Sood, Hiteshi Oza, Abhay Despande, and Tanmay Belsare from Symbiosis Statistical Institute, Pune, Maharashtra; Sarthak Khandelwal and Aditi Dandawate from Medi-Caps University, Indore, Madhya Pradesh; and Akalbir Singh Chadha and Yashowardhan Shinde from International Institute of Information Technology, Hinjewadi, Pune, Maharashtra, for the valuable research support provided to us on several chapters.

We wish to thank Padmashri Prof. Amitav Mallik, Founding Trustee, Pune International Centre and Head, Energy Environment and Climate Change (EECC) & Founder, Climate Collective, Pune, and Mr. Dinanath Kholkar, Vice President and Global Head, Partner Ecosystem and Alliances, Tata Consultancy Services, for critically evaluating and writing the foreword for the book. We would also like to extend our deepest gratitude and appreciation to our affiliations—Tata Consultancy Services, Society for Data Science, Royal Statistical Society, and IEEE, Pune Section, for constant motivation.

Our sustained effort towards writing this book would not have been possible without the consistent support from our family members. Dr. Vijay Sharma, Kshitij Sharma, Teesta Banerjee, Adrija De, and Adrish De deserve a special mention for their patience and constant encouragement to work hard. "Thank You" for being our backbone during this journey of authoring this intellectual asset.

Needless to say, any errors that remain are our own.

Neha Sharma
Prithwis Kumar De

Contents

About the Authors

Neha Sharma is a data science crusader who advocates its application for achieving sustainable goals, solving societal, governmental and business problems as well as promotes the use of open data. She has more than 22 years of experience and presently working with Tata Consultancy Services and is a Founder Secretary, Society for Data Science. Prior to this, she has worked as Director of premier Institute of Pune that run post-graduation courses like MCA and MBA. She is an alumnus of a premier College of Engineering and Technology, Bhubaneshwar and completed her Ph.D. from prestigious Indian Institute of Technology, Dhanbad. She is Senior IEEE member, Secretary—IEEE Pune Section and ACM Distinguished Speaker. She is an astute academician and has organized several national and international conferences and published several research papers. She is the recipient of "Best Ph.D. Thesis Award" and "Best Paper Presenter at International Conference Award" at National Level. She is a well-known figure among the IT circle and well sought over for her sound knowledge and professional skills. Neha Sharma has been instrumental in integrating teaching with the current needs of the Industry and steering students towards their bright future.

Prithwis Kumar De is a Data, Analytics and Artificial Intelligence (AI) expert with over 24 years of experience and a Sustainability Evangelist. He is currently working with Tata Consultancy Services. Prior to this, he has worked with Accenture, CRISIL, Ernst and Young and the Indian Institute of Foreign Trade. He has managed and supported clients across industries such as BFSI, retail, CPG, life sciences and healthcare, telecom, among others, to drive accelerated business outcomes through data and analytics. He holds a Ph.D. in Econometrics (Applied Statistics). Dr. De is a Fellow of the Royal Statistical Society, UK. He has been granted Chartered Statistician (CStat) status, the highest professional honour in Statistics by the Royal Statistical Society. Dr. Prithwis is a member of the American Statistical Association (PStat®), USA. He is a distinguished speaker and has published several papers in professional and refereed journals, books and newspapers. With a passion for sustainability, Prithwis enjoys working on a wide range of topics related to ESG, climate change, business, technology and AI to address the rapidly emerging needs in this space. He opines that data, analytics and technology play a critical role in creating data-driven environment-friendly solutions towards achieving net-zero emissions target. Through this book, he would like to provide the future generations a decarbonized environment and a better tomorrow.

Chapter 1
Climate Change and AI in the Financial, Energy, Domestic, and Transport Sectors

1.1 Introduction

Climate change has been a significant concern since the late twentieth century. It is an observed regional as well as a global change in climate due to an increase in the level of carbon dioxide (CO_2) and other GreenHouse Gases (GHGs) in the atmosphere, which is because of anthropogenic activities like the consumption of fossil fuels [1]. Anaerobic (in the absence of oxygen) decomposition of dead organisms containing organic molecules buried in the earth for millions of years results in fossil fuel as an outcome of a natural process [2–4]. Nevertheless, it is still known as a non-renewable source of energy, as its formation takes millions of years, and the rate of consumption is faster than its formation [5, 6]. Carbon is the main content of fossil fuel and is found in three states—volatile (gas), liquid, and non-volatile (solid). The carbon-to-hydrogen ratio is lower in volatile fossil fuels than in liquid and non-volatile fossil fuels. A hydrocarbon gas mixture consisting primarily of methane is an example of volatile fossil fuel, petroleum is an example of liquid fossil fuel, and coal is an example of non-volatile fossil fuel [7].

According to the BP Statistical Review of World Energy 2020, the primary sources of energy in 2019 are oil (33.1%), coal (27.0%), and gas (24.2%), which cater to 84.3% of the energy required from fossil fuels [8]. The remaining 15.7% of energy comes from non-fossil sources like nuclear (4.3%), hydroelectric (6.4%), and renewables (5.0%) [8]. Renewable energy is the energy generated from renewable sources that are carbon–neutral and easily replenishable, and it includes geothermal heat, solar, tidal, wind, wood, waves, and biomass [9, 10]. According to the REN21 2020 report, the Total Final Energy Consumption (TFEC) share of renewable energy is 18.1%, and that of fossil fuel is 79.9% [11]. Both the reports, i.e., BP Statistical Review of World Energy 2020 and REN21 2020, bring to our attention the fact that the ratio of energy consumption from fossil fuel to renewable energy is approximately 80:20, which is alarming.

The combustion of fossil fuels is the primary reason for air pollution that raises a severe threat to the environment and leads to multiple health issues like asthma

N. Sharma and P. K. De, *Towards Net-Zero Targets*, Advances in Sustainability Science and Technology, https://doi.org/10.1007/978-981-19-5244-9_1

and diabetes [12]. Thirty-seven billion tons of CO_2 is produced every year due to the burning of fossil fuels, and a very small portion is absorbed by nature, whereas a major part remains in the atmosphere resulting in air pollution [13, 14]. It results in global warming and ocean acidification, increases the economic burden by 3.3% of GDP worldwide, and leads to the death of 5 million people [15]. This can be directly attributed to the industrial revolutions that began in the seventeenth century and the global rush towards economic development [16]. The Sixth Assessment Report (AR6) of the Intergovernmental Panel on Climate Change (IPCC, 2019), published in 2019, clearly indicates industrial revolution and human activities are the prime reasons for GHG emissions, which mainly include CO_2, CH_4 (methane), O_3 (ozone), and N_2O (nitrous oxide) [17]. However, a global mission to reduce the use of fossil fuels would save 3.6 million lives every year [15].

It is imperative to address these challenges posed by climate change at multiple levels. In this chapter, an attempt has been made to plunge into four key sectors—financial, non-renewable, households, and transport sectors—that are largely responsible for the change in the climate. The next four sections (Sects. 1.2, 1.3, 1.4 and 1.5) introduce these sectors and their impact on the environment. Further, data science and Artificial Intelligence (AI) have been explored to mitigate climate risks. This chapter acts as a vehicle for discussion and deliberations in the following chapters of this book.

1.2 Carbon Emissions, Climate Change, and the Financial Sector

According to the Fossil Fuel Finance Report, 2021, banks have invested around $3.8 trillion in fossil fuel companies between 2016 and 2020, even after signing the Paris Agreement. The three banks that have done the most fossil fuel financing in 2020 are JPMorgan Chase at $51.3 billion, Citi Bank at $48.4 billion, and Bank of America at $42.1 billion [18]. A commitment to be net-zero by 2050 has been made by 60 banks, but according to the report, the promises are weak, as some banks have policies to reserve finance for the companies dealing with fossil fuel, oil, and gas [18]. Many banks such as Bank of America Corp, Barclays Plc, and Morgan Stanley have promised to track and report the carbon emissions resulting from their lending and investments. According to a report by a climate non-profit organization—Climate Disclosure Project (CDP), GHG emissions from financial institutions' investing and lending in carbon emissions producing industries are 700 times higher, on average, than their direct emissions [19].

On the contrary, there are acts of making false or misleading claims about the environmental benefits of a product, service, technology, etc., by the fossil fuel sectors. The shares of companies in such sectors, which eventually lead to extensive carbon emissions and climate change, are sold by the banks. This is an alarming issue that requires some profound changes in the existing investment policies. Therefore, the financial sector has an important role to play in reducing emissions through sustainable finance and responsible investments.

1.2.1 Climate Change and Financial Sector

With a rapid transition to a low-carbon economy, there would be little ground for central banks to rescue the holders of assets in carbon-intensive companies. While banks in financial distress in an ordinary crisis can be resolved, this will be far more difficult in the case of economies that are no longer viable because of climate change. Two main factors by which climate change can affect financial stability are physical and transition risks.

Physical risks are those risks that arise from climate-related hazards and their exposure to human and natural systems. They represent the economic costs and financial losses due to the increasing frequency and severity of climate-related weather events (e.g., storms, floods, or heat waves) and the effects of long-term changes in climate patterns (e.g. ocean acidification, rising sea levels, or changes in precipitation). The losses incurred by firms across different financial portfolios (e.g., loans, equities, and bonds) can make them more fragile.

Transition risks are associated with the unpredictable financial effects resulting from a rapid low-carbon transition, including policy changes, new technological breakthroughs or changes, or limitations in market choices and social values. In particular, a quick transition to lower emissions means that a large fraction of proven fossil fuel reserves cannot be extracted, becoming "stranded assets", and these stranded assets might lose value in the future, with potentially systemic consequences for the financial system. Too rapid a movement towards a low-carbon economy could materially damage financial stability. As climate-related risks are re-evaluated, a wholesale reassessment of prospects, could destabilize markets and lead to a persistent tightening of financial conditions. Moreover, the value added of many other economic sectors dependent on fossil fuel companies will probably be impacted indirectly by transition risks. For instance, the automobile industry may be strongly affected as technologies, prices, and individual preferences evolve. Assessing how the entire value chain of many sectors could be affected by sudden ups and downs in the supply of fossil fuels is particularly challenging [20].

1.2.2 AI and the Finance Sector

Environmental, Social, and Governance (ESG) is a popular approach to evaluate companies and countries on how far advanced they are with sustainability. ESG analysis has become an increasingly important part of the investment process. Investors are incorporating ESG data into the investment process to gain a fuller understanding of the companies in which they invest. Sustainability goals metrics are not commonly part of mandatory financial reporting, though companies increasingly make disclosures in their annual or a stand-alone sustainability reports. Numerous institutions, such as the Sustainability Accounting Standards Board (SASB), the Global Reporting Initiative (GRI), and the Task Force on Climate-related Financial Disclosures (TCFD), are working to form standards and define materiality to facilitate the incorporation of these factors into the investment process [21].

Investors continue to struggle with questions about ESG data collection and interpretation. The absence of standardized ESG datasets and reporting methodologies makes it difficult for issuers to disclose meaningful information on sustainability. Data may not be widely available or manually collected by analysts, leaving data providers little choice but to produce subjective qualitative assessments. This approach means that different ESG research companies and data providers use their own, often inconsistent methodologies to generate ESG scores. However, AI capabilities like Natural Language Processing (NLP) techniques can be used to collect, clean, and analyze both structured and unstructured data. ESG analysis using AI can give more insights into the kind of investments that are being made and information about their safety to the environment [21].

AI can also be used to keep greenwashing practices by companies in check by financial institutions [22]. Greenwashing is a situation when the impact or the risk of the company's sustainability profile or financial ESG products are not adequately disclosed to the investors. Banks can leverage the online content available in social media, print media, portals, etc., to build AI solutions to verify claims by corporates towards sustainability.

Financial institutions can contribute towards sustainable development by carrying out responsible investments in companies that aim at reducing energy consumption and carbon emissions. ML modeling can be carried out to predict the consequences related to climate change due to the investment by banks in different sectors. The banking sector can explore various ML algorithms, and data science approaches to assess different sectors and analyze their carbon footprint. This may help the banks to take a prudent decision to invest in low-carbon assets.

1.2.3 Conclusions

Investments by the banks and financial sectors can pose serious threats to the environment and lead to climate change. It is difficult to identify which sectors and stocks will create more risk. Most of the current financial system focuses on quarterly or yearly performance, but this fails to take into account the long-term impacts of changes to the climate. Banks and financial institutions can play a vital role in tackling climate change by keeping a check on two factors mentioned below:

1. Reallocate investment in stock markets towards less polluting sectors.
2. Thrust carbon-intensive sectors to develop and implement greener technologies.

A lot can be achieved by integrating AI in the financial sector to devise new methods and even improve the existing ones. AI can also keep greenwashing practices in control and enhance ESG analysis to achieve the net-zero goal by reducing overall carbon emissions.

1.3 Carbon Emissions, Climate Change, and the Energy Sector

Most of the energy consumed is generated from fossil fuels including coal, oil, and natural gas, and takes thousands of years for formation. These sources of energy cannot be reused or renewed and are finite in nature. The dependence on fossil fuels to meet energy demands has increased in recent decades. Since the industrial revolution, fossil fuels have become a dominant source of energy globally. As per a study in 2019, 91.04% of India's energy demand is met by fossil fuels [23].

1.3.1 Climate Change and the Energy Sector

Figure 1.1 highlights the global energy-related carbon emissions for two decades, i.e., 2000–2020 [24]. It represents a consistent rise in carbon emissions during the entire period, calling for immediate interventions to address the same. However, there is a dip in carbon emissions in 2020, which might be due to the COVID-19 pandemic that forced several lockdowns and put a halt to economic activities across the world. This reduction in CO_2 emissions has been the largest since 2010 than the previous record reduction of 0.4 billion tons in 2009.

The burning of fossil fuels emits vast amounts of CO_2 and other harmful GHGs, thus becoming a huge contributor to climate change and the greenhouse effect. As GHG emissions from human activities increase, they build up in the atmosphere and warm the planet. Moreover, because these GHGs have the potential to stay in the atmosphere for tens and hundreds of years after being released, their impact on the climate is long-lasting, affecting current and future generations. Energy efficiency often plays out quietly in the background but has been effective in making huge strides in lowering GHG emissions over the years. Greater investment in energy efficiency can potentially help mitigate the impacts of climate change.

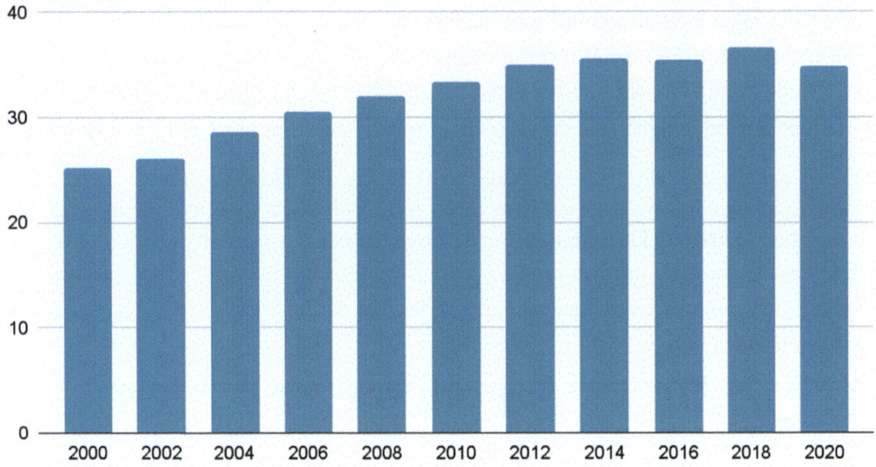

Fig. 1.1 Carbon (CO_2) Emissions from 2000 to 2020 in billion tons

1.3.2 Artificial Intelligence (AI) and the Energy Sector

AI and energy are far more interrelated than accounted for. AI is poised to transform the entire energy sector by helping overcome the variable and uncertain nature of energy and accelerating the adoption of renewables [25].

A. **Areas in which AI is being used in the Energy Sector**

AI is present in the field of intelligent networking of electricity generators and consumers across sector boundaries. With the increasing decentralization and digitalization of the power grid, it is becoming more difficult to manage a large number of grid participants and keep the grid in balance. This requires evaluating and analyzing a flood of data. AI helps process this data as quickly and efficiently as possible [26].

Another application area is smart grids, a network that transports not only electricity but also data. Especially with an increasing number of volatile power generation plants such as solar and wind, it is becoming more and more important for power generation plants to react intelligently to consumption (and vice versa). AI can help evaluate, analyze, and control the data of the various participants (consumers, producers, storage facilities) connected to each other via the grid. In addition, AI can stabilize the power grid by, for example, detecting anomalies in the generation, consumption, or transmission in near real time, and then developing suitable solutions. Initial research projects in this field, such as at the Fraunhofer Institute, are already underway.

AI has the potential to cut energy waste, lower energy costs, and facilitate as well as accelerate the use of clean, renewable energy sources in power grids worldwide. AI can also improve the planning, operation, and control of power systems. Thus, AI technologies are closely tied to the ability to provide clean and cheap energy that is essential for development [27].

AI technology can harness large volumes of data from diverse sources to come up with solutions to problems. It has the potential to increase not only global productivity but also lower overall emissions of carbon and other potent GHG, according to a recent report published by Microsoft in association with PwC [28]. AI for environmental applications can boost global GDP by 3.1–4.4%, as well as reduce global GHG emissions by 1.5–4.0% by 2030 [28]. These AI technologies can help the industry optimize energy management by better monitoring and distributing energy resources and providing a predictive forecast. They can also be used to visualize simulations, improve decision-making, reduce operational costs, and manage and extend the life cycle of physical assets.

For oil and gas companies to sustain themselves in the future they must find ways to dramatically reduce, if not totally eliminate, their output of CO_2 and other GHGs. AI technology could provide some solutions to help the energy industry accomplish that staggeringly difficult goal [29].

B. **AI's role in reducing carbon emissions in the energy sector**

There are some areas in which the implementation of AI methods can substantially improve the effectiveness of current solutions in the energy sector and support faster integration of renewable energy sources. ML algorithms can identify patterns and insights within large datasets to improve prediction accuracy. This would allow energy companies to:

i. **Anticipate energy demand**

Increased accuracy in short-term demand forecasting can assist energy companies in improving production decision-making, enhancing dispatch efficiency, and reducing required operating reserves. AI-based models may help reduce utility bills by optimizing home solar and battery systems for end-consumers while also shifting loads to periods when electric grids are less congested. Mountain View-based start-up Bidgely Inc. uses AI to provide granular visibility and analytics into households' energy generation and consumption [29]. Through it, users can identify which electric appliances are being used and when, as well as how efficiently they are performing, enabling the offering of real-time energy-saving plans, rate programs, and rebates.

ii. **Forecast weather conditions**

Balancing different sources of energy in a renewable portfolio is no easy task, given their dependency on atmospheric conditions. Self-learning weather models, coupled with large-scale processing of satellite imagery and real-time measurements, can significantly reduce renewable energy generation costs.

iii. **Fault prediction**

It has been one of the major applications of AI in the energy sector, along with real-time maintenance and identification of ideal maintenance schedules. In an industry where equipment failure is common, with potentially significant consequences, AI combined with appropriate sensors can be useful to monitor equipment and detect failures before they happen, thus saving resources, money, time, and lives.

iv. **Improving natural gas burner control with process optimization**

Profire Energy Inc., a technology company that provides solutions for industrial combustion appliances for the Oil and Gas (O&G) industry, participated in two phases of Amii's Reducing Emissions through Machine Intelligence (REMI) program. The team identified and explored a high-value and low-risk opportunity to improve burner control by using ML technology to detect out-of-tune burners within the program. These burners waste natural gas and increase GHG emissions—as Profire's install base is 75,000 systems, the savings potential of such a technology is significant. In addition, if this technology works, it could also be applied to an estimated 1.7 million oil wells in North America. There is no objective or cost-effective way to continually monitor and identify an out-of-tune burner for natural gas heaters in the current market. Field operators must make manual observations of burner performance based on flame color, shape, noise, manual combustion analysis, and other field observations. This is costly and time-consuming and results in suboptimal burner management, wasted natural gas, and increased GHG emissions [30, 31].

Profire's proposed solution involved installing additional low-cost sensors to the burner control system and streaming the data to a cloud server, where an ML-based application would automatically predict out-of-tune burners. Field operators would then be notified to schedule priority maintenance to tune the burner, thereby reducing wasted natural gas and lowering GHG emissions.

1.3.3 Conclusions

AI holds considerable potential to improve power generation, transmission, distribution, and consumption. The energy sector in both emerging markets and advanced economies continues to face multiple challenges in terms of efficiency, transparency, affordability, and the integration of renewable energy sources in power systems. The potential applications of AI in the power sector are multiple and varied, and there is a need to educate the AI industry more deeply on the aspects of the power sector. AI plays a critical role in reducing carbon emissions by predicting energy demand, weather conditions, faults in the assets, and optimizing and controlling the process of a natural gas burner.

1.4 Carbon Emissions, Climate Change, and the Domestic Sector

The domestic or the household sector, from their usage, results in 72% of the global GHG emissions [32]. Thus, this sector is one of the main inhibitors in accomplishing the goal of the Paris Agreement, which aims to keep the global average temperatures below 1.5 °C [32, 33]. However, the possible contributions and position of households in climate policies are neither well understood nor do they receive sufficiently high priority in current climate policy strategies. Indeed, very steep reductions in emissions are needed if the global community is to meet the goals of the Paris Agreement, which translate into a reduction of emissions from 40 Giga Tons of CO_2 in 2020 to 5 Giga Tons in 2050 and eventually reach a level of "net-zero" by 2100 [32]. In essence, this means that emissions must be cut in half every decade from now until the middle of the century, after which they must cross the line and undertake so-called negative emissions policies until the end of the century [34].

1.4.1 Leading Contributors to Carbon Emissions from the Domestic Sector

The baseline carbon footprint (CF) assessments of the households participating in the HOPE (Household Preferences for reducing GHG Emissions) study found the mobility sector to be the most significant contributor to a median household's footprint. It may be noted here that the emissions of GHG are accounted for as CO_2 equivalent (CO_2e). Figure 1.2 reveals that the mobility segment accounts for 34% of the CF, followed by food (30%), housing (21%), and others (15%). The median value of CO_2e per household member is calculated as 2.9 tons.

The chart in Fig. 1.2 above includes four different areas that are contributing to carbon emissions from the domestic sector, which are:

1. **Housing:** It includes various appliances that a household uses on a daily basis, like a fan, cooler, or air conditioner in summers or heaters in winters. Additionally, microwaves, television, water pump, and geyser, which consume electricity and even generate carbon emissions, come under this area.
2. **Mobility:** It incorporates all the means of transportation used to commute from one place to another by members of that house, be it a car, a two-wheeler, or an airplane. It includes both electric and fuel-based vehicles.
3. **Food:** This segment talks about the food intake of a family. The carbon emissions in this sector also include the amount of energy and resources used to cultivate

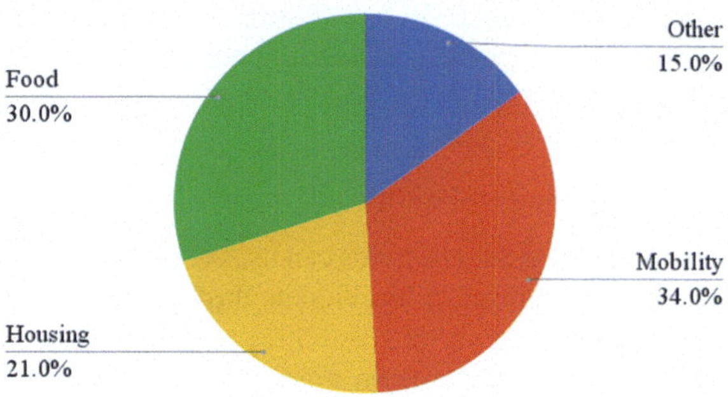

Fig. 1.2 Percentage of Carbon Dioxide Equivalent of various sectors towards climate change

the food, cook the food, and the waste that is generated from the agricultural produce or dishes.

4. **Other:** All the other tasks that require energy and lead to carbon emissions come under this segment, like disposal and management of all kinds of wastes (degradable, non-biodegradable, and e-waste) from a household.

1.4.2 Variation in Carbon Emissions on the Basis of Age of Human Beings

Household carbon footprints are not static as they vary according to situations such as place of living, choice of vehicle—fuel or electricity, the standard of living, etc. Similarly, key life events like graduation from a university, having children, purchasing a home, retiring, or suffering from any medical ailment can significantly impact emissions trajectories [35]. The estimation of an age-specific profile of CO_2 emissions leads to relevant and novel empirical results. Such a profile differs from typical age patterns of overall consumption and production. The shape of the CO_2 emissions profile depends on demographic factors as well as the technology in use and the structure of the economy. It is an informative result for analyzing the life cycle of CO_2 emissions at an individual level to assess the effect of macro-demographic changes on emissions. For instance, it is reported that incidences of illness and poor health are generally related to lower home energy use and reduced mobility. However, it is also linked to higher electricity consumption when income and other demographic attributes are controlled for. As per the study by Zagheni, Metric Tons of CO_2 emissions are highest when an individual attains the approximate average age of 62.

1.4.3 Role of AI in Tackling Climate Change Due to the Domestic Sector

AI can reduce GHG emissions by 5–10% globally, which would translate to 2.6–5.3 fewer Giga Tons by 2030, as per the estimates presented by Morgan et al. [36]. AI can also optimize logistics, reduce the materials required to build products, and hence reduce carbon emissions. Following are some examples where AI is being used to tackle the climate crisis:

A. **Monitoring emissions**

Governments and households can use AI and data engineering to automatically track the carbon footprint of individuals throughout their emissions cycle. The data can be collected from various activities such as travel, food habits, housing, lifestyle, hobbies, etc., and also from every part of the value chain, including materials and components suppliers, transporters, and even downstream users of their products. Besides, AI can exploit data from new sources such as satellites. ML can help by calculating an individual's carbon footprint and identifying small changes that can be made to reduce the emissions—like frequent use of public transport or minimizing the electricity usage in the houses. Adding up to such individual actions can create a big cumulative effect. An ML algorithm can be devised to track and monitor the time, location, source, and destination of people who share the same set of requirements and help them travel together, thereby reducing carbon emissions.

B. **Predicting emissions**

Predictive AI can forecast future emissions across a household's carbon footprint in relation to current reduction efforts, new carbon reduction methodologies, and future demand. As a result, they can set, adjust, and achieve reduction targets more accurately.

C. **Reducing emissions**

Prescriptive AI and optimization can improve efficiency in production, transportation, etc., by providing detailed insights into every aspect of the value chain, thereby reducing carbon emissions and cutting costs. The study by Lacoste et al. proposed an ML Emissions Calculator, which helps to estimate carbon footprint and ultimately reduce carbon emissions [37].

D. **Build better electricity systems**

ML can make electricity systems more efficient by reducing the number of wasted journeys and increasing vehicle efficiency. AI could also reduce car usage through the deployment of shared and autonomous vehicles. Electricity systems are overflowing with data, but too little is being done to take advantage of this information. ML could help by predicting electricity generation and requirements, which will further help to incorporate renewable resources (like wind energy) in supply chains throughout the nation and even reduce the wastage of electricity.

1.4.4 Conclusions

Households will undoubtedly shape future emissions pathways in complex and meaningful ways. They are influential stakeholders in climate policies with substantial control over emissions profiles in their respective households and countries. Yet their potential for achieving such drastic reductions in carbon footprints is uncertain.

1.5 Carbon Emissions and Climate Change Due to the Transport Sector

A large proportion of GHG emissions can be attributed to transportation, specifically from the burning of gasoline, diesel, and other fuels derived from petroleum. Reducing the dependency of transport on petroleum is one of the most direct connections between the issues of climate change and energy security. The use of petroleum fuels in transportation also leads to emissions of particulate matter, sulfur dioxide (which forms sulfate aerosols and ultimately leads to acid rain), and substances that are precursors to photochemical smog (nitrogen oxides—NO_x and carbon monoxide—CO) and various forms of pollution in freshwater and marine systems. Hence, efforts to reduce GHG emissions in the transport sector will also confer other benefits to the environment and public health [38]. Emissions from the transport sector are one of the major contributors to climate change—about 14% of annual emissions (including non-CO_2 gases) and around a quarter of CO_2 emissions from burning fossil fuels [39].

Figure 1.3 shows the data on the worldwide carbon emissions according to the mode of transport. The maximum amount of emissions is caused by the most common mode of transport for daily commuting across the world—the passenger cars [40]. It also helps to identify the area that needs major changes to reduce carbon emissions, which in the case below is the passenger car segment.

It is essential to use public transportation since it is more fuel efficient when considered holistically. When all individuals use their own vehicles (passenger cars) to reach their destination, they are individually contributing a similar amount of carbon to global warming. For example, suppose a single passenger car emits X units of carbon, and all passenger cars are assumed to emit X units for the same duration. In that case, having "n" passenger cars travel to the same destination could make the total emissions X*n units, whereas only X units will be emitted when using public transport.

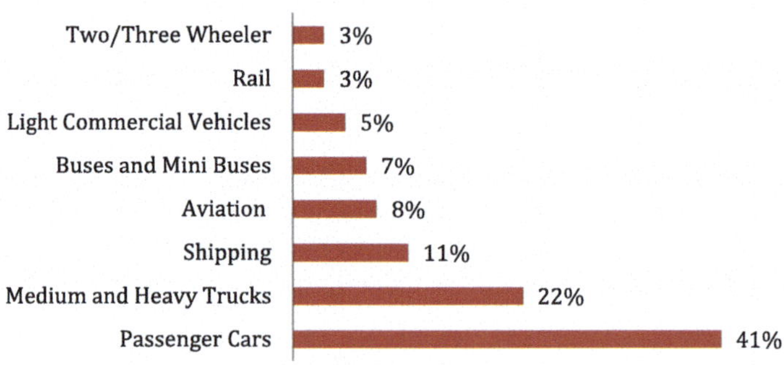

Fig. 1.3 Distribution of CO_2 emissions produced by the transportation sector worldwide in 2020

1.5.1 AI in the Transport Sector

In recent years, AI techniques have become an integral part of realizing smart transportation. The complex interactions among roadways, transportation traffic, environmental elements, and traffic crashes have been explored using an improved deep learning model. Statistical ML algorithms have also found their way into supporting smart transportation. In the past, the duration of the traffic incident on the highway has been estimated with high accuracy, using various ML algorithms like neighborhood components analysis, the Bayesian optimization algorithm, and a random forest model [41]. Besides, there are pieces of evidence showing the profuse use of statistical ML for transportation. For example, a Naive Bayesian classifier is used to identify public transit commuters' travel patterns based on both the smartcard and survey data sampled from commuters [41].

1.5.2 Use of AI to Reduce Climate Change Due to the Transport Sector

This section presents a few interesting AI solutions to tackle climate risks arising from the transport sector.

A. **Expediting resolution of delays and traffic jams reduction**

A lot of fuel is wasted due to traffic jams, which eventually leads to significant carbon emissions. ML-powered transportation platforms crawl through historical data of already traveled delivery routes and generate critical insights to boost fleet productivity and reduce costs. For example, one of the delivery trucks needs to travel through points A, B, and C to reach a customer's destination. Referring to historical data, ML can benchmark the time taken to get point B from A and point C from B. If the duration to reach any of these points exceeds the already set threshold, it will immediately trigger alerts and help transportation stakeholders take quick action. This eliminates the chances of further delays.

B. **Transport efficiency**

ML algorithms are also used for predictive analytics of machinery and accurate assembly of components, which can result in an improved design or functionality of vehicle features and a longer product lifetime. This predictive maintenance process is a standard method to prevent failures of machinery components and is effective in using less material (and energy) than in regular production and maintenance processes. Further enhancements in energy consumption through ML tools can be achieved in the field of autonomous vehicles by avoiding traffic congestion or improving the efficiency of platooning effects (close distance of driving through autonomous vehicles while vehicles are allowed to brake and accelerate simultaneously). Fully connected vehicles will pollute less through a most efficient way of driving, and the possibility of broken machinery (and coherently more energy emissions through additional productions) can be minimized through a better understanding of all data, referring to the concept of digital twins.

C. **Electric vehicles**

Electric vehicles contribute to reducing carbon emissions as they use batteries to run. ML algorithms can assist the recharging process, as failures in charging can be prevented by early detection of anomalies, or charging patterns can support the adequate installation of new charging stations. Equally, the production process of batteries, as well as the maintenance, can be implemented more efficiently and rapidly, as failures can be detected easily. The electric car manufacturer Tesla is dominating the market of electric and autonomous vehicles with the help of two technological advantages: (1) full self-driving car computer technology, which uses two in-house AI processors that are more powerful than the market standards and (2) in-house battery production centers that let them produce electric cars are more cost-efficient. ML methods support this cheaper production process via continuous learning progress from performance data such as cell temperature, charge currents, and swelling levels.

D. **To make car-pooling and shared cabs easier**

An ML model could be devised that is capable of tracking the departing time of people, the location they are leaving from, and the destination they want to go to. The model can deduce people who tend to travel on the same route simultaneously and suggest a set of people to take one single means of transportation instead of each person taking up their individual vehicle.

1.5.3 Conclusions

Technologies have been problem-solvers for businesses, whether it is retail, banking, insurance, healthcare, or even sports. The transport industry has been a major contributor to the movement of people and goods across various geographical regions. It plays a significant role in the supply chain management system, where goods are transferred from one place to another. The industry plays a vital role in moving goods to the right place at the right time in a logistics chain. In order to reap the complete benefit from business investment, technologies like ML, AI, and IoT, among others, have been used by governments and organizations. There are a lot of scopes where using ML and AI, the problem of carbon emissions and climate change can be reduced in the transport sector.

1.6 Carbon Emissions and Climate Change Due to AI

Modernization can only be achieved by innovation and consistent hard work. Advancing from the stone age to the high-tech society, humans have gained a lot of milestones. One of those milestones is the idea of AI, where machines are prepared to behave similarly as that to human beings. Internet and AI systems have changed the lifestyle of an individual drastically. However, the development of these systems comes with a cost. They impact the environment adversely, which might have some unusual imprints on human civilization. AI holds one of the major stakes in climate change, and with time, this proportion has been increasing exponentially. Sometimes, it is the hardware that causes issues, and at other times, it is human error and inconsistencies that lead to extra carbon emissions. Strubell et al. focus on the issues like retraining of ML models, an unidentified bug in the code, lack of proper planning for the system, etc., and propose to include financial and environmental costs of training ML models while building them. This can help fellow researchers to decide on a particular algorithm or model to employ, on the basis of its cost to the environment and economy [42].

On the contrary, AI has also helped mitigate the effect and causes of climate change. For instance, the study by Schmidt et al. facilitated the visualization of the impact of floods in areas where it is prone to occur so that effective measures can be taken prior to such kind of impact of climate change [43]. To understand the growing interest among the researchers towards "Artificial Intelligence and Climate Change", a "Google Trends" for the last 18 years (i.e., since 2004) has been referred. A time graph shown in Fig. 1.4 clearly shows an increasing interest in the areas mentioned above, which has intensified since 2016.

1.6.1 Positive Contribution of AI Towards Climate Change

AI contributes towards eliminating carbon emissions by preventing, monitoring, predicting, and mitigating the causes of carbon emissions.

A. **Preventing Carbon Emissions**

With efficient methods and minimum resources, AI has proved to be a concrete supporter of preventing carbon emissions. Joseph Doyle et al. propose a server

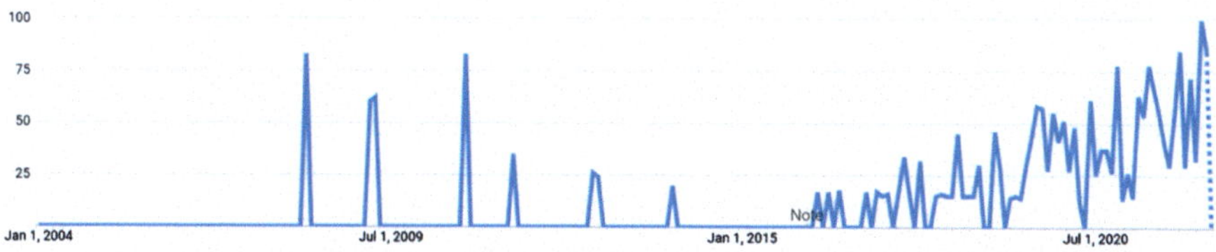

Fig. 1.4 Google Trend for the search term "Artificial Intelligence and Climate Change"

selection algorithm to control the trade-off between carbon emissions and Quality of Service (QoS) [44].

B. **Monitoring and Predicting Carbon Emissions**

Monitoring carbon emissions is one of the major requirements in the modern world to get analysis and take appropriate actions. The study by Nguyen et al. proposed a smart water system named "Autoflow" to monitor and predict the carbon footprint of water consumption. At the core, Autoflow utilizes the power of Artificial Neural Network (ANN) to build and provide services like showing the carbon emissions level and suggesting options that lead to a reduction in carbon emissions [45]. Cai et al. suggest a model to monitor the carbon emissions at Beijing Winter Olympic Games 2022 [46]. Carbon emissions calculators also help flawlessly in monitoring the carbon footprint of systems. For example, a study by Ajufo et al. recommends a personal carbon emissions calculator that helps estimate the carbon footprint of transportation [47]. They also built a mobile application named "TCTracker" using GPS and AI functionality.

C. **Mitigating Carbon Emissions**

As human civilization advances, the need to maintain a proper balance in carbon levels in the atmosphere arises. AI helps greatly in fulfilling those needs. A study by Mao et al. exhibits a reduction of approximately 34,896 tons of CO_2e from conventional buses to electric vehicles [48].

1.6.2 Negative Contribution of AI Towards Climate Change

While AI helps control carbon emissions, it is also a source to spread it to some extent. There are various practices in the AI sector that leave a carbon footprint, and it is essential to discuss the same. Several scientists identified the other side of AI on climate and the reasons which lead to an increase in carbon emissions. The study by Roy et al. bifurcates AI in two parts, viz, Green AI and Red AI [49]. The green AI deals with the type of AI systems which are more efficient and produce less carbon emissions, whereas the red AI focuses on the kind of AI systems which aim towards achieving maximum accuracy without keeping into consideration the efficiency due to which the carbon footprint of the systems increases. Following are some of the factors that contribute to an increased stake of AI in climate change:

A. **Computational Power**

With the growth in the size of algorithms and data, the demand for compatible systems to store and process also increased. Thus, the tech community also advanced in the direction of making processors like AMD Ryzen 5 5600G, whose base clock can reach up to 4.4 GHz. While some processors depend on renewable sources of energy, many others do not. The high processing requires the equivalent amount of electricity, and thus even the wastage of a second can produce much carbon emissions. Moreover, many data centers consume colossal amounts of energy in an inconsistent manner [50].

B. Networking Devices

Modern devices are powered by the fifth generation (5G) network which can download and upload massive amounts of data in just a few seconds. A network such as this requires a dense number of connections. Technologies such as millimeter waves, smart cells, massive Mimo, etc., are major components of 5G networks [51]. The 5G network thus uses more energy than the previous generation, and if such a requirement is fulfilled by non-renewable sources of energy, it leads to more carbon emissions. Furthermore, a study by Durgam et al. shows the adverse effect of mobile tower radiation on birds in the nearby areas of Bijapur district, Chattisgarh, India [52].

C. ML Model Performance

In order to achieve more accuracy, sometimes researchers neglect the cost required for it. So et al. showed that Neural Architecture Search (NAS) achieved a BLEU score of 29.7, which is just a 0.1 increase over the previous state-of-the-art algorithm [53]. This score was acquired at the cost of 150 K USD. A study by Patterson et al. mentions that a typical ML model needs less energy in training than is required in inferencing [54]. Companies like Amazon and NVIDIA have 90% of their ML workload on inference processing. In their study, Patterson et al. also highlighted the carbon footprint of ML models such as T5, Meena, GShard, Switch Transformer, and GPT-3. Furthermore, they also suggest some efficiency improvements in the models.

D. Overall Energy Supply of Selected Companies

In modern times, most of the companies are not green as their energy supply is still coming from non-renewable sources. Gary et al. present a report containing the energy distribution of different companies such as Amazon, Adobe, Google, Microsoft, etc., as shown in Table 1.1 [55].

It is evident from Table 1.1 above that among the companies considered, Apple has the cleanest energy consumption, followed by Google and Adobe. However, it can be seen that Amazon and Oracle have more than 50% of their energy coming from non-renewable sources of energy. With time, these companies have reduced exhaustible energy consumption, but more room for improvement is still required [55].

Table 1.1 Energy distribution of different companies (in Percent)

Name	Clean energy	Natural gas	Coal	Nuclear energy	Others
Google	56	14	15	10	5
Amazon	17	24	30	26	3
Adobe	23	37	23	11	6
Apple	83	4	5	5	3
Oracle	8	26	36	25	5

E. **Internet Traffic**

In modern times where every application requires the Internet for its work, concern about the contribution of the Internet to carbon emissions has increased. Social media applications such as WhatsApp, Instagram, Discord, etc., have taken deep roots within the life of almost every individual. If companies providing such services get even a minimal amount of their energy distribution from non-renewable energy sources, it could create a vast difference in their carbon footprint. Even sending emails results in carbon emissions. Griffiths, in her study, mentioned that 3.7% of the global carbon emission is attributed to the usage of gadgets and the Internet, which is equivalent to the carbon footprint of the airline industry [56].

1.6.3 Conclusions

AI systems have a significant impact on controlling the levels of carbon dioxide in the atmosphere. They are continuously experimented to come up with new and innovative methods to reduce carbon emissions. They help mitigate carbon emissions in almost every domain and are considered to be more efficient than human work. However, along with a positive side, its negative side shows us the side effects of inconsistent or reckless usage of these systems. Green AI can make systems more and more efficient to reduce their carbon footprint and help to achieve net-zero. Furthermore, AI can play a crucial role in limiting carbon emissions by introducing innovative and smart methods.

References

1. Nesbit Jeff (2013). Settled science. Climate change is for real. US News. https://www.usnews.com/news/blogs/at-the-edge/2013/09/24/settled-science
2. Motoaki Sato (1990) Thermochemistry of fossil fuel formation, U.S. Geological Survey, Special Publication No.2, MS 959, Reston, Virginia 22092, U.S.A
3. Schmidt-Rohr K (2015) Why combustions are always exothermic, yielding about 418 kJ per Mole of O2. J Chem Educ 92:2094–2099. http://dx.doi.org/https://doi.org/10.1021/acs.jchemed.5b00333
4. Paul Mann, Lisa Gahagan, Mark B Gordon (1990–1999) Tectonic setting of the world's giant oil and gas fields. In Michel T Halbouty (ed) Giant oil and gas fields of the decade. Tulsa, Okla: American Association of Petroleum Geologists, p. 50
5. Miller G, Spoolman S (2007) Environmental science: problems. Cengage Learning, Connections and Solutions
6. Ahuja, Satinder (2015) Food, energy, and water: the chemistry connection. Elsevier. ISBN 978-0-12-800374-9
7. Fossil fuel. ScienceDaily. https://www.sciencedaily.com/terms/fossil_fuel.htm
8. BP Statistical Review of World Energy 2020.69 Edition, BP. 2020
9. Ellabban, Omar, Abu-Rub, Haitham, Blaabjerg, Frede (2014) Renewable energy resources: current status, future prospects and their enabling technology. Renew Sustain Energy Rev 39:748–764 [749]. https://doi.org/10.1016/j.rser.2014.07.113

10. Harvey, Chelsea, Heikkinen, Niina (23 March 2018) Congress says biomass is carbon neutral but scientists disagree—using wood as fuel source could actually increase CO_2 emissions. Scientific American
11. Renewables 2020 Global Status Report—REN21
12. Myllyvirta L (2020) Quantifying the Economic Costs of Air Pollution from Fossil Fuels
13. Ambrose, Jillian (12 April 2020) Carbon emissions from fossil fuels could fall by 2.5bn tonnes in 2020. The Guardian. ISSN 0261–3077
14. What Are Greenhouse Gases?. US Department of Energy
15. Zhang, Sharon. Air Pollution Is Killing More People Than Smoking—and Fossil Fuels Are Largely to Blame. Pacific Standard
16. Readfearn, Graham (27 September 2013) Planet Ozblog badge Previous Blog home IPCC climate change report by numbers. The Guardian. Retrieved 27 September 2013
17. Climate Change 2013: The Physical Science Basis
18. Banking on Climate Chaos, Fossil Fuel Finance Report, 2021. https://www.ran.org/wp-content/uploads/2021/03/Banking-on-Climate-Chaos-2021.pdf
19. The Time to Green Finance, CDP Financial Services Disclosure Report 2020. https://cdn.cdp.net/cdp-production/cms/reports/documents/000/005/741/original/CDP-Financial-Services-Disclosure-Report-2020.pdf?1619537981
20. Bolton P, Despres M, Pereira Da Silva LA, Samama F, Svartzman R (2020) The green swan. Central banking and financial stability in the age of climate change. January 2020. https://www.bis.org/publ/othp31.pdf
21. Kozlov M, Kulkarni A, Li X, Ho D (2021) Using AI to Tackle the ESG Data Challenge, Worldquant. https://www.worldquant.com/ideas/using-ai-to-tackle-the-esg-data-challenge/#:~:text=AI%20can%20become%20the%20key,opportunities%20to%20create%20new%20ones
22. Kishan S (2021) Using artificial intelligence used to find corporate greenwashers green insight. https://www.bloomberg.com/news/articles/2021-08-11/using-artificial-intelligence-used-to-find-corporate-greenwashers-green-insight
23. https://aeee.in/decoding-energy-and-climate-change/
24. Hannah Ritchie, Max Roser, Pablo Rosado (2020) CO_2 and greenhouse gas emissions. Published online at OurWorldInData.org. https://ourworldindata.org/co2-and-other-greenhouse-gas-emissions
25. https://sifted.eu/articles/ai-energy-transform/
26. https://www.next-kraftwerke.com/knowledge/artificial-intelligence
27. https://www.ifc.org/wps/wcm/connect/bd3a196d-a88f-45af-bbc6-e0b00790fba8/EMCompass_Note_81-05-web.pdf?MOD=AJPERES&CVID=n72pj5g#:~:text=Artificial%20intelligence%2C%20or%20AI%2C%20has,and%20control%20of%20power%20systems
28. How AI can enable a sustainable future (2019) A Joint Report by Microsoft and PWC
29. https://www.bidgely.com/
30. Magill J (2021) Oil, gas companies deploy AI in the fight to reduce carbon emissions. Sustainability, Forbes. https://www.forbes.com/sites/jimmagill/2021/03/28/oil-gas-companies-deploy-ai-in-the-fight-to-reduce-carbon-emissions/?sh=11db0041761e
31. Kube R, Ayotte B. (2021) 4 Ways AI is Helping Companies Reduce Greenhouse Gas Emissions. amii. https://www.amii.ca/latest-from-amii/4-ways-ai-ghg-emissions/
32. IPCC (2018) Summary for policymakers. In: Global Warming of 1.5°C. "An IPCC Special Report on the impacts of global warming of 1.5°C above pre-industrial levels and related global greenhouse gas emission pathways, in the context of strengthening the global response to the threat of climate change, sustainable development, and efforts to eradicate poverty" [Masson-Delmotte V, P Zhai, H-O Pörtner, D Roberts J Skea, PR Shukla, A Pirani, W Moufouma-Okia, C Péan, R Pidcock, S Connors, JBR Matthews, Y Chen, X Zhou, MI Gomis, E Lonnoy, T Maycock, M Tignor, T Waterfield (eds)]. In Press. https://www.ipcc.ch/sr15/
33. Dubois G, Sovacool B, Aall C, Nilsson M, Barbier C, Herrmann A, Bruyère S, Andersson C, Skold B, Nadaud F, Dorner F, Moberg KR, Ceron JP, Fischer H, Amelung D, Baltruszewicz

M, Fischer J, Benevise F, Louis VR, Sauerborn R (2019) It starts at home? Climate policies targeting household consumption and behavioral decisions are key to low-carbon futures. Energy Res Soc Sci 52:144–158. https://doi.org/10.1016/j.erss.2019.02.001
34. https://www.myclimate.org/information/faq/faq-detail/what-are-negative-emissions/
35. Zagheni E (2011) The leverage of demographic dynamics on carbon dioxide emissions: does age structure matter? Demography 48(1):371–399. https://doi.org/10.1007/s13524-010-0004-1
36. Morgan L (2021) AI carbon footprint: Helping and hurting the environment. Sustainability in Business. https://searchenterpriseai.techtarget.com/feature/AI-carbon-footprint-Helping-and-hurting-the-environment
37. Lacoste A, Luccioni A, Schmidt V, Dandres T (2019) Quantifying the carbon emissions of machine learning. http://arxiv.org/abs/1910.09700
38. 13 Transportation. National Research Council (2010) National Academies of Sciences, Engineering, and Medicine. 2010. Advancing the Science of Climate Change. Washington, DC: The National Academies Press. https://doi.org/10.17226/12782
39. Wang S, Ge M (2019) Everything you need to know about the fastest-growing source of global emissions: transport. World Resources Institute. https://www.wri.org/insights/everything-you-need-know-about-fastest-growing-source-global-emissions-transport
40. Distribution of carbon dioxide emissions produced by the transportation sector worldwide in 2020, by subsector (2021) Statista. https://www.statista.com/statistics/1185535/transport-carbon-dioxide-emissions-breakdown/
41. Tizghadam A, Khazaei H, Moghaddam MHY, Hassan Y (2019) Machine learning in transportation. J Adv Transp Hindawi. https://doi.org/10.1155/2019/4359785
42. Strubell E, Ganesh A, McCallum A (2019) Energy and policy considerations for deep learning in NLP. arXiv preprint arXiv:1906.02243
43. Schmidt V, Luccioni A, Mukkavilli KS, Balasooriya N, Sankaran K, Chayes J, Bengio Y (2019) Visualizing the consequences of climate change using cycle-consistent adversarial networks. https://doi.org/10.48550/arXiv.1905.03709
44. Doyle J, O'Mahony D, Shorten R (2011) Server selection for carbon emission control. GreenNets '11: Proceedings of the 2nd ACM SIGCOMM workshop on Green networking. Association for Computing Machinery, New York, NY, USA. pp 1–6. https://doi.org/10.1145/2018536.2018538
45. Nguyen KA, Sahin O, Stewart RA, Zhang H (2017) Smart technologies in reducing carbon emission: artificial intelligence and smart water meter. ICMLC 2017: Proceedings of the 9th International Conference On Machine Learning And Computing. Association for Computing Machinery, New York, NY, USA. pp. 517–522. https://doi.org/10.1145/3055635.3056566
46. Cai Y, Lin Z, Liu L, Wang Y (2018) Demand forecast for winter olympic games food cold chain logistics with carbon emissions consideration. ICIME 2018: Proceedings of the 2018 10th International Conference on Information Management and Engineering. Association for Computing Machinery, New York, NY, USA. pp 177–182. https://doi.org/10.1145/3285957.3285983
47. Ajufo CAM, Bekaroo G (2021) An automated personal carbon footprint calculator for estimating carbon emissions from transportation use. icARTi '21: Proceedings of the International Conference on Artificial Intelligence and its Applications. 12, pp 1–7. https://doi.org/10.1145/3487923.3487935
48. Mao F, Li Z, Zhang K, Zhang D (2020) Impacts on carbon dioxide emissions from the replacement of conventional buses by electric buses. DSIT 2020: Proceedings of the 3rd International Conference on Data Science and Information Technology. pp 175–180. https://doi.org/10.1145/3414274.3414501
49. Schwartz R, Dodge J, Smith NA, Etzioni O (2020) Green AI. Commun ACM 63(12):54–63. https://doi.org/10.1145/3381831
50. Glanz J (2012) Power, pollution and the internet. The New York Times. https://www.nytimes.com/2012/09/23/technology/data-centers-waste-vast-amounts-of-energy-belying-industry-image.html

51. Amy N, Kristen C (2017) Everything you need to know about 5G: millimeter waves, massive MIMO, full duplex, beamforming, and small cells are just a few of the technologies that could enable ultrafast 5G networks. IEEE Spectrum. January 27
52. Durgam DK, Sao S, Singh RK (2017) Effect of mobile tower radiation on birds in Bijapur district, Chhattisgarh. World J Pharm Pharmac Sci 6:1221–1229
53. So D, Le Q, Liang C (2019) The evolved transformer. International Conference on Machine Learning. Proceedings of the 36th International Conference on Machine Learning, PMLR, 97:5877–5886
54. Patterson D, Gonzalez J, Le Q, Liang C, Munguia L, Rothchild D, So D, Texier M, Dean J (2021) Carbon emissions and large neural network training. https://doi.org/10.48550/arXiv.2104.10350
55. Cook G, Lee J, Tsai T, Kongn A, Deans J, Johnson B, Jardim B (2017) Clicking clean: who is winning the race to build a green internet? Technical report, Greenpeace
56. Griffiths S (2020) Why your internet habits are not as clean as you think. Smart Guide to Climate Change. https://www.bbc.com/future/article/20200305-why-your-internet-habits-are-not-as-clean-as-you-think

Chapter 2
Role of Banking Sector in Climate Change—Literature Review and Data Preparation

2.1 Introduction

The world has witnessed tremendous economic progress, and humanity has taken giant steps in its journey through time. However, the side effects of these developments are significant and have an enormous impact on the world. Climate change and global warming are major issues faced by our planet as an outcome of these developments. As per the Inter-Governmental Panel on Climate Change (IPCC), scientific evidence for climate system warming is unequivocal [1]. The rising concentration of GreenHouse Gas (GHG) emissions in the atmosphere have emerged as the primary contributor to climate change. As per data available from the National Aeronautics and Space Administration (NASA), global carbon emissions have risen by 30% and the sea level by over 3.4 mm per year [2, 3]. Also, global temperatures are on the course to increase by 3.5 °C at the end of 2100 [2, 4]. Figure 2.1 depicts that, as per predictions under different policies and pledges, the global temperature is going to increase tremendously in the future [4]. The impacts of these changes have consequences on our weather, oceans, food, and health [1].

The world is at the stage of getting faced with extreme weather events, and only those industries that will allow business models to transition to net-zero policies will be rewarded in future. Climate change also poses longer term risks to financial institutions and high-intensity physical risks. The destruction of coastal areas due to rising sea levels and other ruins caused by floods, wildfires, and cyclones of increasing severity and frequency have immediate and short-term risks. Numerous groups, such as the Task Force on Climate-related Financial Disclosures (TCFD), have highlighted that even leading financial risk analysis frameworks fail to account for climate risks as well as other similar risks [2]. Limiting global warming and achieving the 1.5–2 °C goal requires a transformation of the structure of international economic activities on a large scale [1]. The success of any climate change policy relies heavily on its potential effect on the economy [1]. According to the TCFD report published in 2019, the non-financial sectors such as energy, transport, building, and agriculture led the way in 2017 in disclosing regulatory information related to climate change. In

N. Sharma and P. K. De, *Towards Net-Zero Targets*, Advances in Sustainability Science and Technology, https://doi.org/10.1007/978-981-19-5244-9_2

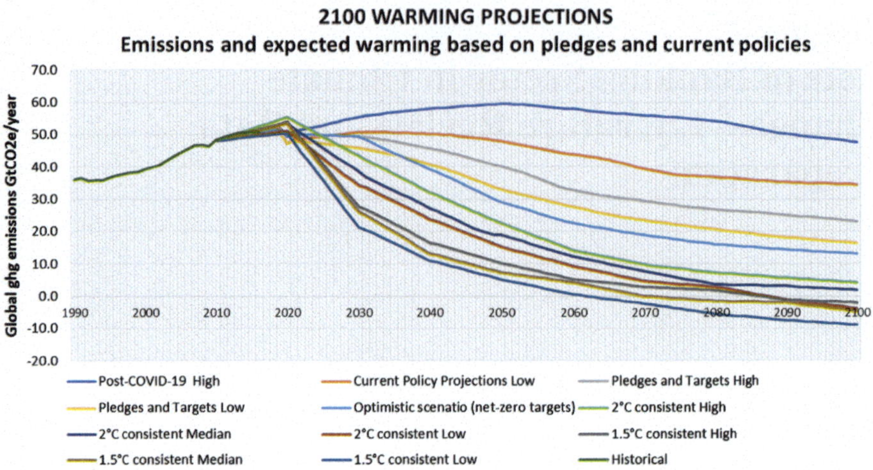

Fig. 2.1 Different policy pathways for Global Warming projection till 2100

contrast, the banking sector is now leading in sharing information [5]. The Year-on-Year growth of rates of disclosure by different sectors for the years 2019 and 2020, respectively, are shown in Tables 2.1 and 2.2.

Table 2.1 indicates that the banking sector led the disclosure rate in almost all the TCFD recommendations in 2019, whereas Table 2.2 reflects that the Energy sector led in 2020.

The literature on climate change and its impact on the economy is overwhelming in terms of both the depth of analysis and coverage of the range of complex issues. While a lot of work has been done in the context of developed economies, the literature is still at a nascent stage for developing economies. Policymakers and central bankers are also realizing the growing economic risks of climate change, necessitating changes in the systems for modeling and forecasting macroeconomic variables [6–8]. The literature suggests that climate-related disclosures could lead to an orderly transition to a low-carbon economy if the process helps most investors better evaluate their financial risk exposures [8, 9].

The role of a bank is crucial for economic and social well-being. They are the providers of capital and decision-makers for the sectors to be financed. Nevertheless, they are also exposed to physical risks (such as heatwaves, floods, cyclones, storms, and drought) and transition risks (from financing carbon-centric sector to low-carbon sector) due to climate change. Physical risks include damage to the balance sheets of households, corporates, banks, and insurers triggered by a weather-related event causing financial and macroeconomic instability, whereas the transition risks would consist of the after-effects of the implementation of fiscal policy tools, which could lead to re-pricing of carbon-intensive assets and also trigger adverse supply shocks. Figure 2.2 depicts the physical and transition risks.

The banking industry has been considered to be a non-environmentally sensitive industry due to the non-polluting nature of its business [10, 11]. Banks may not

Table 2.1 Rates of disclosure against each of the TCFD recommendations by sector (TCFD Report 2019, the numbers in the parentheses represent the size of the review population)

Recommendation	Recommended disclosure	Banking (104) (%)	Insurance (147) (%)	Energy (128) (%)	Materials & buildings (213) (%)	Transportation (223) (%)	Agriculture, food & forest (166) (%)	Technology and media (63) (%)	Consumer goods (82) (%)
Governance	a. Board oversight	48	29	38	37	25	22	19	29
	b. Management's role	54	35	32	35	18	26	17	40
Strategy	a. Risks and opportunities	51	39	57	50	39	40	38	50
	b. Impact on organization	55	26	64	65	34	45	25	52
	c. Resilience of strategy	20	12	13	12	5	4	2	6
Risk management	a. Risk ID & assessment processes	52	30	38	41	23	24	24	22
	b. Risk management processes	46	33	42	39	17	26	19	23
	c. Integration into overall risk management	32	16	21	18	11	9	17	21
Metrics and targets	a. Cliamte-related metrics	51	27	49	63	36	45	37	55
	b. Scope 1,2,3 GHG emissions	42	22	39	41	29	26	29	38
	c. Climate-related targets	50	24	45	53	32	30	24	51

Table 2.2 Rates of disclosure against each of the TCFD recommendations by industry (TCFD Report 2020, the numbers in the parentheses represent the size of the review population)

Recommendation	Recommended disclosure	Banking (236) (%)	Insurance (138) (%)	Energy (274) (%)	Materials & buildings (414) (%)	Transportation (162) (%)	Agriculture, food & forest (147) (%)	Technology and media (113) (%)	Consumer goods (167) (%)
Governance	a. Board oversight	23	24	42	26	16	20	6	11
	b. Management's role	24	35	45	31	20	21	8	17
Strategy	a.Risks and opportunities	32	49	65	39	38	41	27	27
	b. Impact on organization	27	21	54	43	30	35	16	25
	c. Resilience of strategy	12	8	15	4	2	1	0	2
Risk management	a. Risk ID & assessment processes	25	36	39	26	15	21	7	16
	b. Risk management processes	23	28	38	26	18	26	11	14
	c. Integration into overall risk management	19	26	30	14	10	14	4	7
Metrics and targets	a.Cliamte-related metrics	27	26	41	43	37	35	23	31
	b. Scope 1,2,3 GHG Emissions	21	19	27	35	27	28	19	24
	c. Climate-related targets	19	22	44	41	35	38	19	29

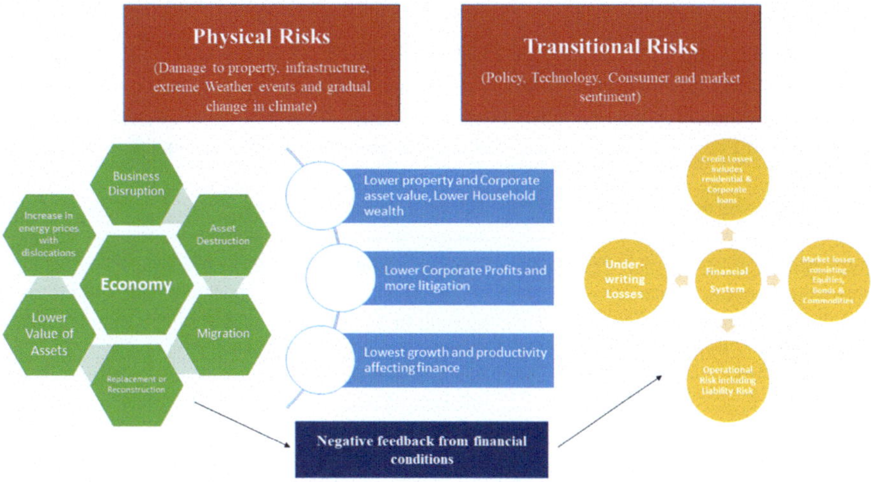

Fig. 2.2 Impacts of physical and transitional risks

be significant polluters themselves, but they might have a relationship with companies/investment portfolios that are carbon-intensive in nature. Therefore, they have an impact on the environment due to their financial intermediary role. Banks can contribute towards sustainable development by providing more capital to products aimed at reducing energy consumption and carbon emissions [12, 13]. They can also declare incentives and benefits to companies having green projects while financing them. The World Business Council for Sustainable Development has identified the financial industry as a critical sector in providing environmental sustainability [13, 14]. The increment in the severity and frequency of climate-related risks directly affects the insurance plan that covers them.

Many initiatives have been started to keep a record of climate change, such as the Global Reporting Initiative, Carbon Disclosure Project, and Greenhouse Gas Protocol, as shown in Table 2.3. Some initiatives that are specific to the finance industry include Principles for Responsible Banking, Task Force on Climate-related Financial Disclosure, Banking Environment Initiative, Principles for Sustainable Banking, Paris Pledge on Coal Financing, and Montréal Carbon Pledge.

While climate change has become a global mission, in this chapter and Chap. 3 make a focused attempt to understand the role of the banking sector and financial institutions in climate change. In these two chapters, the attempt is to define the research objective, identify the data source, perform the statistical analysis, and apply machine learning to mitigate the climate change risk due to carbon emissions by banks and other financial institutions themselves or the establishments directly or indirectly supported by them.

In this chapter, after setting up the context above, an effort has been made in Sect. 2.2 to carry out a detailed literature review to understand the challenge. Section 2.3 presents the identification and preparation of the dataset as well as algorithms used to find the solution and Sect. 2.4 concludes the chapter.

Table 2.3 Climate change initiatives at global level

Sr. No	Name	Year	Objectives
1	United Nations Environment Programme (UNEP) [15]	1972	UNEP, a key environmental authority, advocates and works to implement ecological rules and regulations at a global level. It promotes sustainable development
2	Intergovernmental Panel on Climate Change [1]	1988	The IPCC carries out research and gives a periodical assessment of climate change, its implications, and risks, and propose new mitigation solutions
3	United Nations Environment Programme Finance Initiative (UNEP FI) [16]	1992	UNEP FI was established during the Rio Earth Summit in 1992. It is a collaboration between UNEP and the global financial sector to work for sustainable development in the financial sector. It involves various members, mostly banks, insurers, and investors
4	Global Reporting Initiative [17]	1997	GRI provides sustainability disclosure standards at a global level. They encourage organizations to report climate-related impacts
5	Carbon Disclosure Project [18]	2000	The carbon disclosure project gathers corporate climate change data and persuades companies, investors, and organizations to manage, disclose, and reduce their emissions
6	Green House Gas Protocol [19]	2001	Greenhouse Gas Protocol provides worldwide greenhouse gas accounting rules. It creates global frameworks for measuring and managing GHG emissions from business and public sector activities, value chains, and mitigation initiatives
7	Banking Environment Initiative [20]	2010	Banking Environment Initiative (BEI), started in 2010, is a group of global banks working collectively towards sustainable economic development. It focuses on decarbonizing the global economy
8	Green Climate Fund UNFCC [21]	2010	GCF follows a country-driven approach to support developing countries towards low-emission pathways and adapt to climate change impacts
9	Portfolio Decarbonization Coalition (PDC) [22]	2014	It works to decarbonize the portfolios of institutional investors by withdrawing investments from carbon-intensive companies/projects
10	Montreal Carbon Pledge [23]	2014	It is a global pledge to motivate investment banks to closely observe and decrease the carbon emissions related to their investment portfolios
11	Task Force on Climate-related Financial Disclosure [24]	2015	TCFD aims to create a set of voluntary climate-related financial risk disclosures. The disclosure recommendations are structured around four main areas: governance, strategy, risk management, as well as metrics and targets

(continued)

Table 2.3 (continued)

Sr. No	Name	Year	Objectives
12	Paris Agreement I UNFCC [25]	2016	The Paris Agreement is an international treaty on climate change that works on a 5-year cycle. It was adopted by 196 Parties at COP 21 in Paris on 12 December 2015 and entered into force on 4 November 2016
13	The Network of Central Banks and Supervisors for Greening the Financial System (NGFS) [26]	2017	NGFS works to meet the goals of the Paris agreement and to improve the role of the financial system in low-carbon investments

2.2 Literature Review

The section examines the literature pertaining to the repercussions of climate change due to the banking and financial sector. This section is divided into two subsections: the first one is to present the systematic literature review and the second is to perform a contemporary review of the literature. The systematic literature review helps to find the importance of the topic in present times and the contemporary review puts forward the traditional review of the published articles. At the end, the research objective pertaining to the banking sector, financial sector, and climate change are also pronounced.

2.2.1 Systematic Literature Review

A thorough literature study is conducted in order to assess important work in the fields of climate change, carbon emissions, finance and banking sector, and Machine Learning (ML) solutions for reducing emissions from the banking sector. The information is gathered from three different digital libraries: Springer, IEEE Xplore, and Elsevier. The search is performed using key phrases such as "Climate Change", "Climate Change and Finance", "Banking Sector affecting Climate Change", and "ML model for Climate change due to Banking Sector and Finance". Figure 2.3a, b, and c show that the number of papers published on climate change has been steadily increasing over the previous decade, demonstrating a growing global concern about the issue.

It can be seen from Fig. 2.3a, b and c that Elsevier has the highest number of publications in terms of climate change as compared to IEEE Xplore and Springer. It is evident from the figures that there is an increase in the research published in all the digital libraries on the topic of climate change. However, less research is carried out towards building ML solutions for mitigating climate change due to the banking sector and financial institutions, which creates lots of opportunities for using Data Science as a critical technology to combat climate change in these sectors.

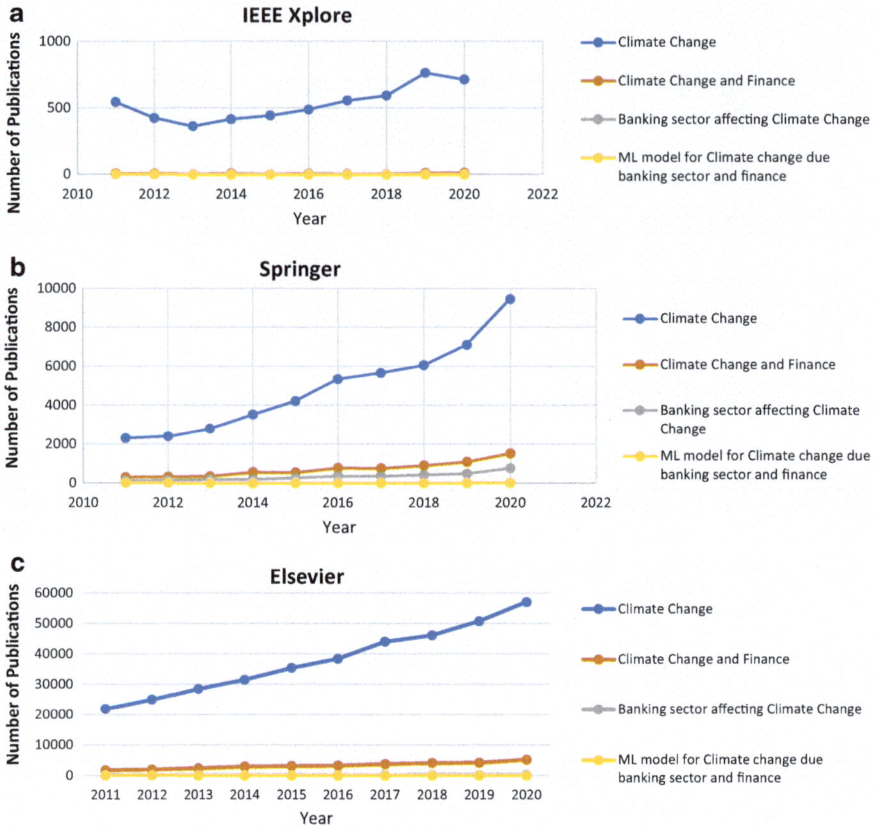

Fig. 2.3 **a** Systematic literature review of IEEE Xplore. **b**. Systematic literature review of Springer. **c**. Systematic literature review of Elsevier

It is important to note that in Fig. 2.3a and c, a few legends are not reflected in the graph as the data points are negligible. It is also observed in Fig. 2.3b that Springer is the only publication that covers all the search phrases that are used to perform a systematic literature review.

2.2.2 *Contemporary Literature Review*

This subsection provides the study of different literature work that has been done on climate change, the banking sector, financial stability, and banks responsible for climate change. Table 2.4 presents the summary of the literature review carried out by different researchers on these topics and highlights the objectives, country of research, dataset, and techniques/algorithm used. The study is mainly performed in four categories as follows:

Table 2.4 Summary of literature review

Domain	Title	Year	Objectives	Country	Dataset	Techniques/ algorithms
Load on Financial Institutions due to natural calamities, physical and transitional risks, policies and economic models	Assessing India's mounting climate losses to Financial Institutions	2019	Risk of extreme weather events to financial institutions	India	None	None
	Climate Change: Macroeconomic Impact and Policy Options for Mitigating Risks	2020	Climate change risk to economies	India	None	None
	Macroeconomic and Financial Policies for Climate Change Mitigation: A Review of the Literature	2019	Role of policy tools in enabling the transition towards a low-carbon economy	None	None	None
	Let's Talk About the Weather: The Impact of Climate Change on Central Banks	2016	The paper looks at how climate change and mitigation initiatives might affect a central bank's capacity to accomplish its monetary and financial stability goals	None	None	None
	Breaking the tragedy of the horizon—climate change and financial stability	2015	Focused on three broad channels through which climate change can affect financial stability	England	None	None
	Multiple regression models to predict the annual energy consumption in the Spanish banking sector	2012	A regression analysis of energy consumption in the Spanish banking sector	Spain	55 Spanish Banks	Multiple Regression

(continued)

Table 2.4 (continued)

Domain	Title	Year	Objectives	Country	Dataset	Techniques/ algorithms
	Using economic model predictive control to design sustainable policies for mitigating climate change	2012	The strengths of the proposed design method are verified and will serve as the foundation for future research on applying systems control theory to design optimal sustainable policies	None	None	Model Predictive Control (MPC), Regional Dynamic Integrated model of Climate and the Economy (RICE model)
	Is there a link between Air Pollution and Economic Growth?	2017	The objective of this study is to investigate the relationship between air pollution and economic growth in Nigeria	Nigeria	World Development Indicators	IBM Watson Analytics
	Carbon tax: Future models and impact	2011	The goal of this study is to describe the most cost-effective energy efficiency finance portfolio for demand-side projects in the context of a carbon price	None	None	None
	'Climate value at risk' of global financial assets	2016	The impact of climate change on the financial sector, especially on asset values	None	None	1. Value at Risk 2. Dynamic Integrated Climate Change model 3. Random parameters and Monte Carlo simulation
Role of central banks in transition to the low carbon economy	Climate change challenges for central banks and financial regulators	2018	The paper outlines the main points of contention and explores future climate change research and policy options	None	None	Macroeconomic modeling such as dynamic stochastic general equilibrium, agent-based models

(continued)

Table 2.4 (continued)

Domain	Title	Year	Objectives	Country	Dataset	Techniques/ algorithms
	The economics of climate change: the Stern review	2007	It focuses on the consequences and risks of unchecked climate change, as well as the costs and benefits of taking action to address it	US	None	None
Carbon Emissions in Banking Sector	Determinants of climate change disclosures in the Turkish banking industry	2019	Examine the amount of voluntary climate change disclosures in the Turkish banking industry, as well as the factors that contribute to that extent	Turkey	24 Banks (2010–2016)	Multiple regression, Fractional regression analysis
	The determinants of voluntary climate change disclosure commitment and quality in the banking industry	2020	The present state of knowledge on the effects of size and country-level characteristics on carbon disclosure is expanded by this research, which finds that size and national context are not independent of carbon disclosure	Global	117 banks (40 developed and developing countries)	Least Square Regression, Multivariate Logit models
	The impact of financial development on carbon emissions: a global perspective	2019	The relationship between financial development and carbon emissions based on a system generalized method of moments and data of 155 countries is analyzed	China	World Bank data, IMF data, 155 Countries dataset	Generalized method of moments, Regression model

(continued)

Table 2.4 (continued)

Domain	Title	Year	Objectives	Country	Dataset	Techniques/ algorithms
	Carbon Emissions Are not just Bad	2019	Correlation between GDP, carbon emissions, and renewable energy production	China	None	Correlation Detection
Role of Task Force on Climate-related Financial Disclosures (TCFD) in Climate Change Mitigation	TCFD: strengthening the foundations of sustainable finance	2020	TCFD Recommendations	Japan	None	None
	Climate Change and Financial Risk: Central banks and financial regulators are starting to factor in climate change	2019	Action that can be taken towards climate change and financial regulators to invest more in renewable sources of energy	None	None	None
	Climate change threatens the stability of the financial system	2019	1. Integrate climate risk into the regulatory and supervisory frameworks 2. Climate Change threatens Financial Stability 3. Focus on Finance Green Industries	USA	None	None

i. Load on financial institutions due to natural calamities, physical and transitional risks, policies, and economic models.
ii. Role of central banks in transition to the low-carbon economy.
iii. Carbon emissions in the banking sector.
iv. Role of Task Force on Climate-related Financial Disclosures (TCFD) in climate change mitigation.

A. **Load on financial institutions due to natural calamities, physical risk, existing policies, and economic models**

Singh, in his study, presented the increasing risks of extreme weather events to the financial institutions in India [27]. The report highlights that there is a strong need for the intervention of the Government of India, various state government bodies, and financial sector regulators to work together to develop a comprehensive strategy for mitigating the impact of this growing risk on individuals, farmers, and businesses. Dilip et al. highlighted the rising risks from climate change to the macroeconomic outlook of economies around the world and also reviewed the available-risk mitigating policy options [8]. Krogstrup et al. provided an overview of the literature on the role of financial policy and shared some views about the policy tools (financial, fiscal, and monetary) in enabling the transition towards a low-carbon economy [28]. Climate-related financial risks are unique, and they have a far-reaching impact of unforeseeable and irreversible nature. They usually stem from short-term actions (NGFS 2019) [28].

Transitioning to a low-carbon economy requires investments to shift from a high-carbon sector to a low-carbon or zero-carbon-intensive sector [9]. Carney states that, if this shift does not take place on a large scale and the policy on carbon emissions is not tightened, then this would result in falling asset prices for those organizations that heavily depend on fossil fuels [29]. Carbon emission-reducing policies can focus on the cap-and-trade system, such as the EU Emission Trading Scheme—ETS) [29]. Aranda et al. presented a regression analysis of energy consumption in the Spanish banking sector [30]. The authors have considered the data from 55 banks and applied 3 predictive models for energy consumption for the whole banking sector, branches with low winter climate severity and branches with high winter climate severity, respectively [30]. The paper concludes that the first model allows for the detection of weak bank branches and can be used as a tool to propose energy-saving measures to reduce the bank branch's energy consumption. Models 2 and 3 are not validated and are expected to achieve a better adjustment using the actual energy consumption [30].

Chu et al. described an approach where Economic Model Predictive Control is applied to a Regional Dynamic Integrated model of Climate and the Economy (RICE) to design sustainable policies that achieve sustainable levels of GHG emissions while minimizing the impact on the economy [31]. The results obtained show the strength of the proposed design approach and form the basis for future research on using systems control theory to design optimal sustainable policies. They have used feedback control theory to design sustainable policies for climate change and global warming [31]. The simulations are based on two different scenarios, the "BAU"

(Business As Usual) and the "Optimal" scenario. The BAU scenario ignores environmental damage and assumes that no action against climate change is adopted and serves as a baseline scenario for comparing different policy designs [31].

John et al. studied the relationship between air pollution and economic growth in Nigeria using the IBM Watson Analytics approach to analyze the data [32]. A smart city framework was developed for informed decision-making and analysis of air pollution. The paper discusses various other analyses done by using Kuznets curve hypothesis. A hypothetical smart city network has been developed, and a detailed analysis has been done, taking into account the different contributors to air pollution (CO_2 emissions from transport, residences, methane, etc.) [32]. To address the challenges of climate change, Bolt et al. state that policies should be implemented as (1) price-based measures, e.g., carbon tax and emission trading schemes, (2) Provision of education and information, and (3) Imposition of standards. They have discussed the carbon tax, carbon trading, and excess energy rates in South Africa [33]. Dietz et al. discussed the utilization of Aggregated Integrated Assessment Models (IAMs) (DICE model structure, random parameters, and Monte Carlo simulation and 2 °C mitigation scenario) to acquire a first gauge of the environment climate Value at Risk (VaR) [34].

B. **Role of central banks in transition to the low carbon economy**

Campiglio et al. analyzed the role of central banks and financial regulators in promoting a better understanding of climate-related financial risks [35]. The transition to a low-carbon economy is a "tragedy of the horizon" as stated by Mark Carney, Governor of the Bank of England, in 2015 [2]. These transition risks would mean fossil fuel reserves being written off the balance sheet of banks and devaluation of the company that owns them. This could trigger a collapse of the financial system. The authors have compared the long-term and short-term low-carbon transition. Emphasis has been given to the disclosure of climate risk among investors, banking portfolios, and business models [35]. Central banks should incorporate climate-related criteria in assessing whether an asset should be eligible for purchase as part of their portfolio management. Climate change is the greatest and the wildest ranging market failure has ever seen [36]. A smooth low-carbon transition will require the implementation of a comprehensive set of policies that involves the collaboration of central banks and financial regulators.

C. **Carbon emissions in the banking sector**

Caby et al. identified the voluntary climate change disclosure in the banking industry and the quality of that disclosure by applying ordinary least square regression and multivariate logit analysis to the data of 177 banks from 40 developed countries around the world [10]. They concluded that the size of the bank and national characteristics drive carbon disclosure compliance [10]. Besides, they extensively addressed the relationship between Corporate Social Responsibility (CSR) and Environmental Social Governance (ESG).

Jiang et al. studied the relationship between financial development and carbon emissions based on a system generalized method of moments by analyzing the data from 155 countries. The empirical results indicated that financial development could significantly increase carbon emissions from a global perspective. The analysis done by the authors regarding emerging markets and developing countries also reached the same conclusion. However, the results indicated that for developed countries, the effect of financial development on carbon emissions is insignificant [37]. Chen et al. analyzed data from the International Energy Organization and the World Bank to find that the developing countries are the main reasons for carbon emissions and the impacts of carbon emissions are both positive and negative. Maintaining a dynamic balance of carbon emissions will contribute to the development of the economy [38]. Kılıç et al. investigated the extent of voluntary climate change disclosure in the Turkish banking industry and found that the bank size, profitability, bank age, and listing status have a significant positive impact on the extent of climate change disclosure [13].

D. **Role of task force on climate-related financial disclosures (TCFD) in climate change mitigation**

In his report, Mark Carney, Governor, Bank of England, has focused on the Task Force on Climate-related Financial Disclosures (TCFD) to strengthen the foundation of sustainable finance [2]. He proposed four different actions to be taken under TCFD to overcome risk and create opportunities, as mentioned below:

1. To build the amount and nature of revelations by sharing best practices.
2. To refine the TCFD revelation suggestions to those that investors consider most helpful.
3. To spread information on the best way to gauge and utilize data on essential strengths to oversee hazards and acknowledge openings.
4. To consider how resource proprietors could best unveil their portfolios that are suited for the progress to net-zero targets.

Grippa et al. carried out an extensive study on climate change affecting the financial system, the economic impact of climate change, and the extent of damage that is going to depend on the policy choices [39]. The authors have summarized that the low-carbon economy might accelerate the cost of energy sources, leading to an opportunity for technological innovation. Financial sectors have a pivotal role to play in this transformation. Gelzinis et al. suggested a few steps that regulators must adopt to integrate climate risk into their regulatory and supervisory frameworks and focus on financing the green industries to accomplish net-zero outflows by 2050 [40]. The authors have also proposed strategies and technology to offset climate-related risks and maintain financial stability.

Finally, after conducting a rigorous literature review, the objective of the research is defined as—the application of machine learning to predict climate change consequences due to investments in fossil fuel sectors by banks and financial institutions.

2.3 Materials and Methods

Climate change is one of the most significant challenges in today's world. Mitigation of the adverse effect of climate change requires a large-scale transition towards a low-carbon economy. This section addresses the data preparation process, which involves data source identification, description of the data, and different algorithms used in the current and the next chapters.

2.3.1 Data Source Identification

In order to address the problem statement stated above, there is a need for qualitative and quantitative information that is relevant, comparable, and up to date. The quantitative data is made up of two major categories as follows:

1. The first category deals with the investments made by the banking sector in different companies in the fossil fuel sector.
2. The second category includes a dataset of various consequences of fossil fuel extraction activities, like GHG emissions, change in mean sea level, change in annual mean surface temperature, and carbon emissions from gaseous, solid, and liquid fuel combustion.

The Rainforest Action Network (RAN) website has been referred to gather data regarding the banks that are financing the companies in the fossil fuel sector [41]. RAN, founded in 1985, is a non-profit environmental organization based out of San Francisco, United States. It works with a mission to protect the earth's rainforests and support the rights of its inhabitants through education, grassroots organization, and nonviolent direct action. It throws challenges to the corporates by means of various campaigns and partnerships with other non-governmental organizations, indigenous forest communities, and student groups [41]. The "Banking on Climate Chaos (2021)" report has been referred to collect relevant and updated data for the research [42]. The report includes financing data (lending and underwriting of debt and equity issuances) from the world's 60 biggest banks for the entire fossil fuel sector and top companies in specific sectors [42]. Therefore, the dataset of the top 60 banks that are investing in fossil fuel sectors is created, which includes:

1. Tar Sands,
2. Coal Power,
3. Coal Mining,
4. Arctic Oil and Gas,
5. Offshore Oil and Gas,
6. Fracked Oil and Gas, and
7. Liquefied Natural Gas (LNG).

In each of the above sectors, the investments made by the banks lead to some major consequences which affect the climate adversely. These seven sectors are used to predict future investments and the consequences like carbon emissions, methane emissions, temperature rise, etc. For example:

1. Tar Sands—Global temperature, mean surface temperature, and CO_2 emission.
2. Coal Power—Methane emission, global temperature, mean surface temperature, and CO_2 emission.
3. Coal Mining—Methane emission, global temperature, mean surface temperature, and CO_2 emission.
4. Arctic Oil and Gas—Methane emission, global temperature, mean surface temperature, and CO_2 emission.
5. Offshore Oil and Gas—Methane emission, global temperature, mean surface temperature, and CO_2 emission.
6. Fracked Oil and Gas—Methane emission, global temperature, mean surface temperature, and CO_2 emission.
7. Liquefied Natural Gas (LNG) —Global temperature, mean surface temperature, and CO_2 emission.

2.3.2 Data Description and Attributes

Hence, to predict climate change consequences due to investments in fossil fuel sectors by banks and financial institutions, two separate datasets have been created by.

1. combining the data of all the seven fossil fuel sectors and
2. deriving climate change consequences-related data from different reputed organizations like NASA and International Monetary Fund (IMF).

The above dataset is prepared in excel. Further, Python programming language and Jupyter notebook have been used to analyze and build the machine learning model. Few libraries, such as Pandas, NumPy, Seaborn, and Matplotlib, are used. Pandas library is used, to import data from various file formats like CSV and Excel, and for data analysis. Various mathematical operations on arrays of the dataset are performed with the help of the NumPy library. While the Seaborn library is utilized for data visualization and exploratory data analysis, Matplotlib is applied to plot various 2D and 3D graphs.

The below code shows the first step of importing the required python libraries:

```
import pandas as pd
import numpy as np
import seaborn as sns
import matplotlib.pyplot as plt
import matplotlib.mlab as mlab
import matplotlib
plt.style.use('ggplot')
from matplotlib.pyplot import figure
```

To know the size of the dataset in the number of rows and columns, the following code snippet is used:

```
[In] df.shape
[Out] (5, 8)
```

It is evident from the code above that there is a dearth of sufficient datasets due to the less research work carried out in this field. To know the column name, data type, and non-null value, the following code is used:

```
[In] df.info()
[Out] <class 'pandas.core.frame.DataFrame'>
RangeIndex: 5 entries, 0 to 4
Data columns (total 15 columns):
```

#	Column	Non-Null Count	Dtype
---	------	--------------	-----
0	Year	5 non-null	int64
1	Methane_Emission_Oil & Gas	5 non-null	float64
2	Methane_Emission_Coal_Mines	5 non-null	float64
3	Min_Arctic_Sea_Ice_Extent	5 non-null	float64
4	Global_Temp	5 non-null	float64
5	Mean_Surface_Temperature	5 non-null	float64
6	Mean_Sea_Level	5 non-null	float64
7	CO2_Avg_ppm	5 non-null	float64
8	Tar Sands	5 non-null	float64
9	Coal Power	5 non-null	float64
10	Coal Mining	5 non-null	float64
11	LNG	5 non-null	float64
12	Arctic Oil & Gas	5 non-null	float64
13	Offshore Oil & Gas	5 non-null	float64
14	Fracked Oil & Gas	5 non-null	float64

```
dtypes: float64(14), int64(1)
memory usage: 728.0 bytes
```

The dataset has two types of data—int64 and float64. All the data values are numeric with no NULL values in the dataset.

2.3.3 Data Preparation

This section describes the datasets related to investment in 07 fossil fuel sectors and consequences in terms of climate change.

Data regarding investments in seven fossil fuel sectors

The dataset from RAN has quantitative details of banks financing the fossil fuel sector from 2016 to 2020 and also investments in over 2300 fossil fuel companies worldwide, totaling $3.805 trillion [42].

Following is a detailed illustration of the dataset from 2016 to 2020:

1. For all ***fossil fuel investments*** comprising 60 banks, 2300 fossil fuel companies, and 07 sectors, it is observed that JPMorgan Chase, Citi, Wells Fargo, the Bank of America, and Royal Bank of Canada (RBC) are the five top banks that are investing the most, whereas Sumi TRUST and Credit Mutuel are having the least investments.
2. Tar sands oil is one of the most carbon-intensive forms of energy and substituting conventional oil with it increases global warming emissions by 20%. ***Tar Sands dataset*** consists of data from 50 banks. It is found that banks financing for top 30 tar sands production companies and 05 key tar sands pipeline companies totaling $117.135 billion, out of which Toronto-Dominion Bank (TD), RBC, JPMorgan Chase, Canadian Imperial Bank of Commerce (CIBC), and the Bank of Montreal are the top investors in this sector. In contrast, investment by banks like Crédit Mutuel, Danske Bank, DZ Bank, National Australia Bank (NAB), Nordea, Rabobank, Sberbank, Shinhan, the State Bank of India, and SuMi TRUST is almost zero.
3. Data for 50 banks are included in ***Arctic Oil and Gas dataset***. The data reflects the details of banks that are financing for top 30 arctic oil and gas companies resulting in a total investment of $31.199 billion. JPMorgan Chase, Industrial and Commercial Bank of China (ICBC), China Minsheng Bank, Sberbank, and Citi Bank are the top investors, while banks like CIBC, Crédit Mutuel, Danske Bank, Lloyds, NAB, Rabobank, Shinhan, SuMi TRUST, Truist, and U.S. Bank have almost zero investment. For certain banks in some particular year, the investment in this sector is nil.
4. Data for 51 banks are present in ***Offshore Oil and Gas dataset.*** The Oil and gas industry generally operates along a coastline or in open ocean waters. It is evident from the dataset that banks financing for the top 30 offshore oil and gas companies have an overall investment of $327.817 billion. Banque National de Paris (BNP) Paribas, JPMorgan Chase, Citi, the Bank of America, and Hongkong and Shanghai Banking Corporation (HSBC) are the topmost investors in this sector. On the other hand, banks like the Bank of Montreal, Crédit Mutuel, Danske Bank, DZ Bank, NAB, Nordea, SuMi TRUST, Truist, and U.S. Bank have not invested in this sector.
5. Data for 46 banks have been considered for ***Fracked Oil & Gas dataset***. The data clearly shows that the banks financing the top 30 fracked oil companies and 10 key fracked oil and gas pipeline companies have an overall investment of $418.772 billion. Wells Fargo, JPMorgan Chase, Citi, the Bank of America, and Barclays are the topmost investors in this sector, while investments made by Commonwealth Bank, Crédit Mutuel, Danske Bank, DZ Bank, Intesa Sanpaolo, NAB, Nordea, Rabobank, Sberbank, Shinhan, the State Bank of India, SuMi TRUST, UniCredit, and Westpac is zero.

6. Data for 56 banks have been taken in the ***LNG (Liquefied Natural Gas) dataset.*** It is observed that banks financing the top 30 liquefied natural gas import and export companies have a total investment of $120.044 billion. Morgan Stanley, Citi, JPMorgan Chase, the Bank of America, and Sumitomo Mitsui Banking Corporation (SMBC) Group are the topmost investors in this sector, while investment by Rabobank, Shinhan, the State Bank of India, and Truist is nil.
7. Data for 52 banks are taken for the ***Coal Power dataset.*** the Bank of China, ICBC, China International Trust and Investment Corporation (CITIC) Bank, Agricultural Bank of China, and China Construction Bank are the top banks that are investing, with a total investment of $212.174 billion in this sector. In contrast, investment made by the Bank of Montreal, CIBC, Crédit Mutuel, Danske Bank, Internationale Nederlanden Groep (ING), Nordea, Rabobank, and Shinhan is zero.
8. Data for 52 banks have been considered for the ***Coal Mining dataset.*** Industrial Bank, China Construction Bank, Bank of China, Shanghai Pudong Development Bank, and China CITIC Bank are the topmost financers and the total investment in this sector is $119.512 billion, whereas Banque Populaire and Caisse d'Epargne (BPCE)/Natixis, Crédit Mutuel, Danske Bank, Shinhan, SuMi TRUST, Truist, U.S. Bank, and Wells Fargo have made no investment.

The comprehensive view of the total investment made, top 3 investors, and information about the banks that have not invested is depicted in Table 2.5.

Figure 2.4 depicts the investment made by the top 10 banks in the fossil fuel sector for 5 years (2016–2020).

It has been observed from Fig. 2.4 that JPMorgan Chase is the highest investor in the overall fossil fuel sector in all the years totaling $316.735 billion.

Data regarding climate change consequences

Data regarding the change in mean sea level, direct and indirect carbon emissions per million USD, annual mean global surface temperature, annual GHG emissions in metric tons of CO_2 equivalent, and other consequences of fossil fuel activities have been derived from multiple sources and a detailed overview of the data has been provided in Table 2.6.

2.3.4 Algorithms and Evaluation Metrics

In this subsection, the algorithms and the evaluation metrics that are used to predict climate change consequences due to investments in fossil fuel sectors by banks and financial institutions are discussed.

Multivariate Regression—Multivariate Regression (MVR) is a supervised machine learning approach that analyses numerous data variables. It is a variant of multiple regression with a single dependent variable and a large number of independent variables. An attempt is made to anticipate the outcome based on the number

Table 2.5 Top 3 most banks that are investing in seven different fossil fuel sectors with overall investments

Fossil fuel sector	Investments (in billion)	Top investors			Banks with zero investment
		Bank 1	Bank 2	Bank 3	
Tar Sands	$117.135	TD Bank	RBC	JP Morgan	Crédit Mutuel, Danske Bank, DZ Bank, NAB, Nordea, Rabobank, Sberbank, Shinhan, **State Bank of India**, SuMi TRUST
Arctic Oil & Gas	$31.199	JP Morgan	ICBC	China Minsheng Bank	CIBC, Crédit Mutuel, Danske Bank, Lloyds, NAB, Rabobank, Shinhan, SuMi TRUST, Truist, US Bank
Offshore Oil & Gas	$327.817	BNP Paribas	JP Morgan	Citi Bank	Bank of Montreal, Crédit Mutuel, Danske Bank, DZ Bank, NAB, Nordea, SuMi TRUST, Truist, US Bank
Fracked Oil & Gas	$418.772	Wells Fargo	JP Morgan	Citi Bank	Commonwealth Bank, Crédit Mutuel, Danske Bank, DZ Bank, Intesa Sanpaolo, NAB, Nordea, Rabobank, Sberbank, Shinhan, **State Bank of India**, SuMi TRUST, UniCredit, Westpac
LNG	$120.044	Morgan Stanley	Citi Bank	JP Morgan	Rabobank, Shinhan, **State Bank of India**, Truist
Coal Power	$212.174	Bank of China	ICBC	CITIC Bank	Bank of Montreal, CIBC, Crédit Mutuel, Danske Bank, ING, Nordea, Rabobank, Shinhan
Coal Mining	$119.512	Industrial Bank	China Construction Bank	Bank of China	BPCE, Crédit Mutuel, Danske Bank, Shinhan, SuMi TRUST, Truist, US Bank, Wells Fargo

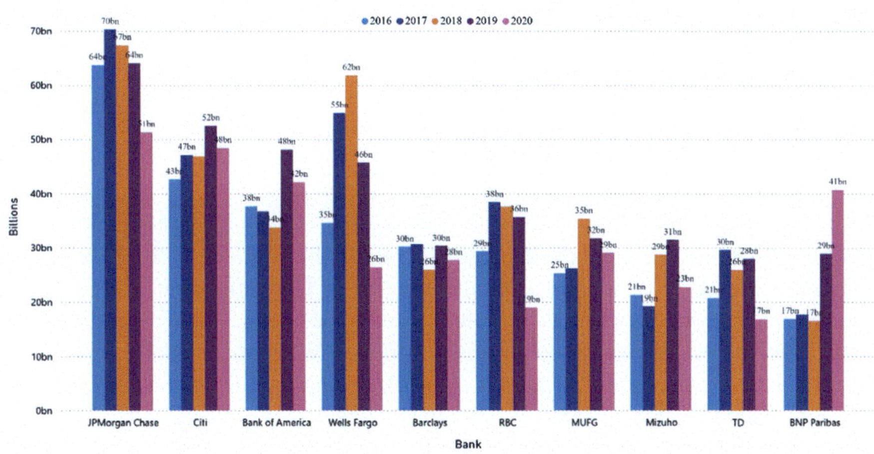

Fig. 2.4 Investment of top 10 banks in overall fossil fuel sectors

Table 2.6 Overview of climate change consequences dataset

S.No	Title	Source	Description	Year
1	Change in Mean Sea Level	IMF Climate Change Dashboard [43]	Dataset of 32,000 rows of worldwide sea level value in different measures like the Andaman Sea, the Arabian Sea, the Atlantic Ocean, the Bay of Bengal, and the Pacific Ocean	1992–2020
2	Annual Global Surface Temperature	IMF Climate Change Dashboard [43]	Dataset of around 229 countries in °C	1961–2019
3	Atmospheric Carbon Dioxide Concentration	IMF Climate Change Dashboard [43]	Dataset of around 1500 rows having monthly and annual atmospheric carbon dioxide emission in ppm and percentage	1958–2021
4	Minimum Arctic Sea Ice Extent	NASA Global Climate Change [44]	Data of minimum arctic sea ice extent in million sq. km	2016–2020
5	Methane Emission Oil and Gas	Global Methane Initiative (GMI) [45]	Worldwide data of methane emission from oil and gas in million metric tons of CO_2	2016–2020
6	Methane Coal Mines	Global Methane Initiative (GMI) [45]	Worldwide data of methane emission from coal mines in million metric tons of CO_2	2016–2020
7	Global Temperature	NASA Global Climate Change [46]	Global temperature in °C	2016–2020

of independent variables. Multivariate regression seeks to discover a formula that can explain how variables respond to changes in others at the same time [47].

$$y = \beta_0 + \beta_1.x_1 + \beta_2.x_2 + .. + \beta_n.x_n$$

where n represents the number of independent variables, $\beta_0 \sim \beta_n$ represents the coefficients and $x_1 \sim x_n$, is the independent variable. The multivariate model helps in understanding and comparing coefficients across the output. Here, a cost function makes multivariate linear regression a better model. When the model differs from observed data, the cost function allows a cost to be applied to samples. The sum of squares of the difference between the anticipated and actual values, divided by twice the length of the dataset, yields this equation. A lower mean squared error indicates an improved performance. The cost is the total of squared mistakes in this case.

$$MSE = \frac{1}{2m} \sum \left(h_\theta(x)^{(i)} - y^i \right)^2$$

Regression Line or Best-Fit Line—A line of best fit is a line that runs across a scatter plot of data points and best reflects the relationship between them. Simple linear regression is a study of two or more independent variables that provide a straight line.

$$b = \overline{y} - m\overline{x}$$

where m = best-fit slope and b = best-fit y-intercept.

Regression line $=(m * x) + b,\ for\ x\ in\ xs$

where xs = input sample on the x-axis.

Mean Squared Error (MSE)—It is the sum of the squared differences between the actual and expected or estimated values. MSE is always positive or larger than zero in value. A value near zero indicates that the estimator/predictor is of higher quality (regression model). The fact that the predictor has an MSE of zero (0) suggests that it is a perfect predictor [48].

$$MSE = \frac{1}{n} \sum\nolimits_{i=1}^{n} \left(yi - \hat{y}i \right)^2$$

where n = Number of samples, yi = Actual target values, and $\hat{y}i$ = Predicted values.

R-Squared—It is the ratio of the Sum of Squares Regression (SSR) to the Sum of Squares Total (SST). The amount of variance explained by the regression line is known as the SSR. The goodness of fit is measured using the R-squared value. The better the regression model, the higher the R-Squared score. The R-Squared is confined between zero and one for the training dataset, but it might turn negative for the test dataset if the Sum of Squared Error (SSE) is bigger than SST. If R-Squared is 1, the model fits the data exactly, with an MSE of 0 as a result [48].

$$R^2 = \frac{SSR}{SST} = \frac{\sum \left(\hat{y}_i - \overline{y} \right)^2}{\Sigma (y_i - \overline{y})^2}$$

where yi = Actual target values, $\hat{y}i$ = Predicted values of y, and $\overline{y}$ = Mean of the y value.

2.4 Conclusions

Climate change is one of the most critical environmental challenges to deal with, and it is becoming more severe day by day. In this chapter, an exhaustive literature review has been carried out regarding the role of the banking sector and financial institutions in achieving the net-zero goal. Besides, climate change-related disclosures in the banking industry, different net-zero strategies, initiatives, economic models, and investments in fossil fuels have been discussed. Data regarding investment by banks in various fossil fuel sectors has been collected from the RAN website, which is a non-profit environmental organization. A dataset regarding various climate change consequences has been collected from NASA and IMF. During the course of data source identification and description, it has been observed that there is a relation between the investment by banks in seven fossil fuel sectors and the consequences of climate change. JPMorgan Chase has emerged as the bank that invests the maximum in the entire fossil fuel sector. In comparison, the State Bank of India has not invested in 3 fossil fuel sectors—tar sands, fracked oil, and gas, and LNG. In the next chapter, exploratory data analysis and machine learning modeling are discussed to predict future investments by banks and the adverse consequences on the environment.

References

1. Stocker TF, Qin D, Plattner G-K, Tignor M, Allen SK, Boschung J, Nauels A, Xia Y, Bex V, Midgley PM (eds) (2013) IPCC, 2013: Climate Change 2013: The Physical Science Basis. Contribution of Working Group I to the Fifth Assessment Report of the Intergovernmental Panel on Climate Change. Cambridge University Press, Cambridge, United Kingdom and New York, NY, USA, p 1535
2. Mark Carney, Governor, Bank of England (2020) TCFD: strengthening the foundations of sustainable finance. SUERF Policy Note, Issue No 122, January 2020
3. NASA: Climate Change and Global Warming (2021). https://climate.nasa.gov/vital-signs/sea-level/
4. Climate Action Tracker (2021). https://climateactiontracker.org/global/temperatures/
5. Michael R. Bloomberg. (2018). Recommendations of the task force on climate-related financial disclosures. Mr. Mark Carney, Chairman Financial Stability Board
6. Nordhaus WD (2017) Projections and Uncertainties about climate change in an Era of minimal climate policies. NBER Working Paper Series, Working Paper 22933
7. Batten S (2018) Climate change and the macroeconomy: a critical review. Staff Working Paper No. 706, Bank of England
8. Dilip A, Kundu S (2020) Climate change: macroeconomic impact and policy options for mitigating risks. RBI Bulletin
9. Batten S, Sowerbutts R, Tanaka M (2016) Let's talk about the weather: the impact of climate change on central banks. Staff Working Paper No. 603, Bank of England
10. Caby J, Ziane Y, Lamarque E (2020) The determinants of voluntary climate change disclosure commitment and quality in the banking industry. Technol Forecast Soc Chang 161:120282. https://doi.org/10.1016/j.techfore.2020.120282
11. Chew BC, Tan LH, Hamid SR (2016) Ethical banking in practice: a closer look at the Co-operative Bank UK PLC. Qualit Res Finan Mark 8(1):70–91
12. Scholtens B (2009) Corporate social responsibility in the international banking industry. J Bus Ethics 86(2):159–175
13. Kılıç M, Kuzey C (2019) Determinants of climate change disclosures in the Turkish banking industry. Intern J Bank Market 37(3):901–926. https://doi.org/10.1108/IJBM-08-2018-0206
14. Buranatrakul T, Swierczek FW (2017) Climate change strategic actions in the international banking industry. Glob Bus Rev 19(1):32–47
15. United Nations Environment Programme (UNEP) (2021). https://www.unep.org/

16. United Nations Environment Programme Finance Initiative—(UNEP FI) (2021). https://www.unepfi.org/
17. Global Reporting Initiative (GRI) (2021). https://www.globalreporting.org/
18. Carbon Disclosure Project (CDP) (2021). https://www.cdp.net/en
19. Greenhouse Gas Protocol (2021). https://ghgprotocol.org/
20. Banking Environment Initiative (BEI) (2021). https://www.cisl.cam.ac.uk/business-action/sustainable-finance/banking-environment-initiative
21. Green Climate Fund (2021). https://www.greenclimate.fund/
22. Portfolio Decarbonization Coalition (PDC) (2021). https://www.unepfi.org/pdc/
23. Montreal Carbon Pledge (2021). https://www.unpri.org/montreal-pledge
24. Task Force on Climate-related Financial Disclosure (TCFD) (2021). https://www.fsb-tcfd.org/
25. Paris Agreement (2021). https://unfccc.int/process-and-meetings/the-paris-agreement/the-paris-agreement
26. The Network of Central Banks and Supervisors for Greening the Financial System (NGFS) (2021). https://www.ngfs.net/en
27. Singh C (2019) Assessing India's mounting climate losses to Financial Institutions
28. Krogstrup S, Oman W (2019) Macroeconomic and financial policies for climate change mitigation: a review of the literature. IMF Working Paper, No. 19/185
29. Carney M (2015) Breaking the tragedy of the horizon—climate change and financial stability, Lloyd's of London, 29 September 2015
30. Aranda A, Ferreira G, Mainar-Toledo MD, Scarpellini S, Sastresa EL (2012) Multiple regression models to predict the annual energy consumption in the Spanish banking sector. Energy Build 49:380–387
31. Chu B, Duncan S, Papachristodoulou A, Hepburn C (2012). Using economic model predictive control to design sustainable policies for mitigating climate change. In 2012 IEEE 51st IEEE Conference on Decision and Control (CDC) (pp. 406–411). IEEE
32. John TM, Ucheaga EG, Badejo JA, Atayero AA (2017) Is there a link between air pollution and economic growth? In 2017 International Conference on Computational Science and Computational Intelligence (CSCI) (pp. 1465–1470). IEEE
33. Bolt GD, Kleingeld M, Van Rensburg JF (2011). Carbon tax: future models and impact. In 2011 Proceedings of the 8th Conference on the Industrial and Commercial Use of Energy (pp. 152–158). IEEE
34. Dietz S, Bowen A, Dixon C, Gradwell P (2016) 'Climate value at risk of global financial assets. Nat Clim Chang 6(7):676–679
35. Campiglio E, Dafermos Y, Monnin P, Ryan-Collins J, Schotten G, Tanaka M (2018) Climate change challenges for central banks and financial regulators. Nat Clim Chang 8(6):462–468
36. Stern N, Stern NH (2007) The economics of climate change: the Stern review. Cambridge University Press
37. Jiang C, Ma X (2019) The impact of financial development on carbon emissions: a global perspective. Sustainability 11(19):5241
38. Chen K, Zhu K, Meng Y, Meng M (2019). Carbon emissions are not just bad. In 2019 4th International Conference on Power and Renewable Energy (ICPRE) (pp. 290–294). IEEE
39. Grippa P, Schmittmann J, Suntheim F (2019) Climate change and financial risk: central banks and financial regulators are starting to factor in climate change. Finance Devel 56(004)
40. Gelzinis G, Steele G (2019) Climate change threatens the stability of the financial system. https://www.americanprogress.org/issues/economy/reports/2019/11/21/477190/climate-change-threatensstability-financial-system
41. Rainforest Action Network (RAN) (2021). https://www.ran.org/bankingonclimatechaos2021/
42. "Banking on Climate Chaos", Fossil Fuel Finance Report, (2021). https://www.ran.org/wp-content/uploads/2021/03/Banking-on-Climate-Chaos-2021.pdf
43. IMF Climate Change Dashboard (2021). https://climatedata.imf.org/pages/re-indicators
44. NASA: Climate Change and Global Warming (2021). https://climate.nasa.gov/vital-signs/arctic-sea-ice/

45. Global Methane Initiative (GMI) (2021). https://www.globalmethane.org/methane-emissions-data.aspx
46. NASA: Climate Change and Global Warming (2021). https://climate.nasa.gov/vital-signs/global-temperature/
47. Multivariate Regression. https://www.mygreatlearning.com/blog/introduction-to-multivariate-regression/
48. MSE & R-Squared. https://vitalflux.com/mean-square-error-r-squared-which-one-to-use/

Chapter 3
Application of Machine Learning to Predict Climate Change Consequences Arising Due to Investments by Banks in Fossil Fuel Sectors

3.1 Introduction

By 2050, millions of people are likely to be affected by drought and water shortages, whereas several major cities of the world are already at risk of coastal flooding [1]. The problem is majorly related to the emissions by the fossil fuel sector and the banks that are financing or investing in them. Therefore, fossil fuel-focused investment is very critical, and banks are expected to position their portfolios to offset the negative impact of climate change. So, the prediction of climate change consequences arising due to investments by financial institutions in the fossil fuel sector is very important.

In Chap. 2, climate change as an outcome of investments by the banking sector and financial institutions has been discussed from various perspectives. Different pieces of literature have been reviewed to understand the load on financial institutions due to natural calamities and physical risk; the role of central banks in the transition to a low-carbon economy; carbon emissions in the banking sector; and the role of Task Force on Climate-related Financial Disclosures (TCFD) in climate change mitigation. The chapter also presented two datasets, i.e., the fossil fuel sector and consequences, along with the algorithms to be used for analysis.

The following section presents the Exploratory Data Analysis (EDA) to describe the dataset and find the relationship between various features. In Sect. 3.3, an analysis of multiple features is performed, and in Sect. 3.4, machine learning models are built and validated. The last section of the chapter presents the conclusion of the research.

3.2 Exploratory Data Analysis (EDA)

EDA is the first step in every data-related project, which helps in understanding different behavior of the data through visual methods. It is a process of data preprocessing data visualization, and it also offers descriptive insights into the data.

N. Sharma and P. K. De, *Towards Net-Zero Targets*, Advances in Sustainability Science and Technology, https://doi.org/10.1007/978-981-19-5244-9_3

3.2.1 *Analysis of Various Fossil Fuel Sectors*

The fossil fuel investment dataset has information related to the investment of 60 banks in 2300 companies under 7 fossil fuel sectors from 2016 to 2020. Figure 3.1 shows the investment of the top 10 banks with respect to total investments in combined fossil fuel sectors. The number above each bar in Fig. 3.1 shows the total investment by respective banks. The different fossil fuel sectors included in Fig. 3.1 are Tar Sands, Arctic Oil and Gas, Offshore Oil and Gas, Fracked Oil and Gas, Liquefied Natural Gas (LNG), Coal Mining, and Coal Power. Since the time Paris Agreement was proposed and adopted, it has been observed that the investment by banks in fossil fuels is $ 3.8 trillion in 5 years. Fracked Oil and Gas, along with Offshore Oil and Gas, are the sectors attracting the most investment by top banks, followed by the Tar Sands sector. Besides, the graph clearly indicates that JP Morgan Chase has invested the most in the fossil fuel sector, followed by Citi Bank.

The following sub-sections present the insights related to the impact on climate due to the activities of seven fossil fuel sectors:

A. **Coal Mining and Coal Power Sector**

Coal is the most abundant source of electricity around the world, currently providing more than 36% of global electricity [1]. While it makes an important contribution to worldwide energy demands, its impact on climate change has been an issue. Due to the lack of modern clean energy, coal power plants are a reliable, affordable, and constant source of power. One of the challenges related to climate change faced by the coal power industry is to develop effective technologies to reduce carbon emissions, especially carbon dioxide. The burning of coal is primarily responsible for the emission of CO (Carbon Monoxide), CO_2 (Carbob dioxide), NO_x (Nitrogen Oxide),

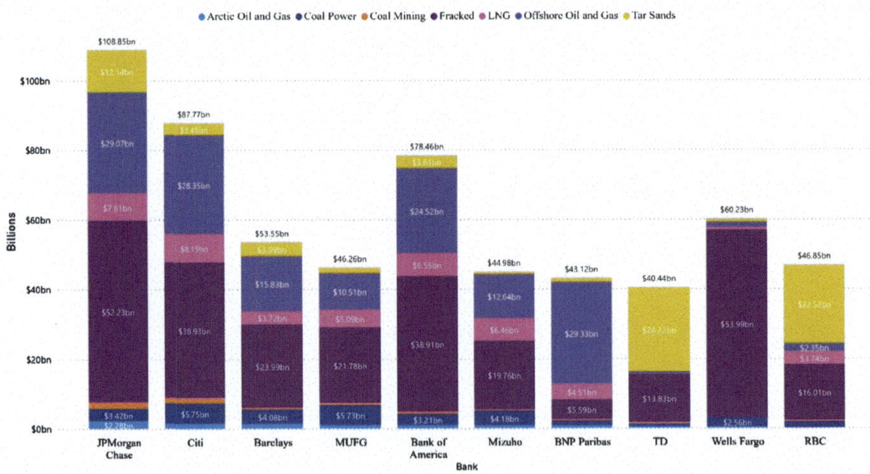

Fig. 3.1 Investment of top 10 banks in different fossil fuel sectors

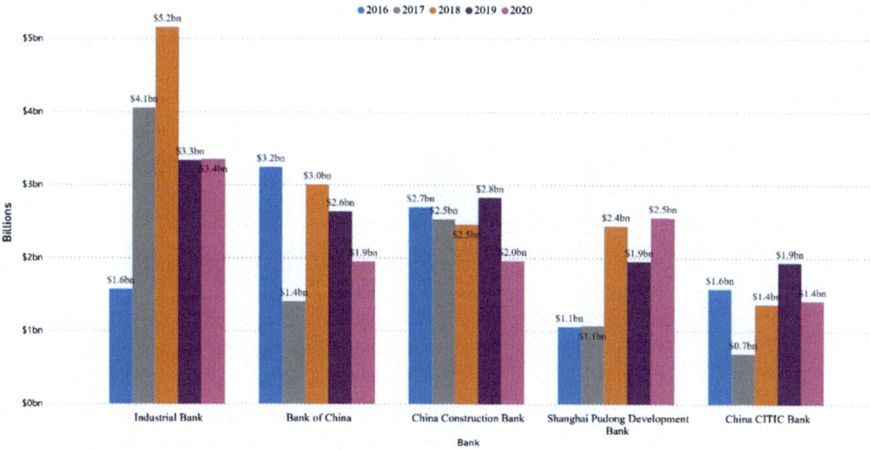

Fig. 3.2 Clustered column chart of top five banks financing the coal mining sector

SO_x (Sulfur Oxide), and CH_4 (Methane). Methane is 21 times more dangerous in the greenhouse effect than CO_2 [2]. The environmental challenges from coal include air pollution, water body pollution, and mining waste disposal. Coal mining also affects water resources by disposing wastewater into nearby water bodies.

The clustered column chart is designed to describe the top 05 banks financing in coal mining and coal power sectors.

From Figs. 3.2 and 3.3, it is obvious that Chinese banks are leading in financing coal mining and coal power sectors. However, in both sectors, it is observed that efforts have been taken to reduce the investments by most banks. Nevertheless, there is an increasing trend of investment observed by Shanghai Pudong Development Bank and China CITIC Bank over the period of 5 years in coal mining and coal power sectors, respectively.

The pie chart is prepared to present the investments made by different countries in these two sectors as shown in Figs. 3.4 and 3.5, respectively. However, the investment made by India in both sectors is negligible.

B. **Tar Sands Sector**

Tar sands, also known as oil sands, are a mix of clay, sand, water, and a thick molasses-like substance called bitumen. Bitumen is made of hydrocarbons and is used to produce gasoline and other petroleum products [3]. Tar sands are not an unlimited source of energy but a relatively secure and stable source. On average, tar sands produce 3–5 times more GreenHouse Gas (GHG) emissions than conventional oil [3]. Besides, tar sands projects are extremely costly and capital intensive. This also raises concerns over its long-time financial usability as the world now focuses on reducing carbon emissions and financing fossil fuel activities [4].

The clustered column chart is designed to describe the top 05 banks financing in tar sands sectors.

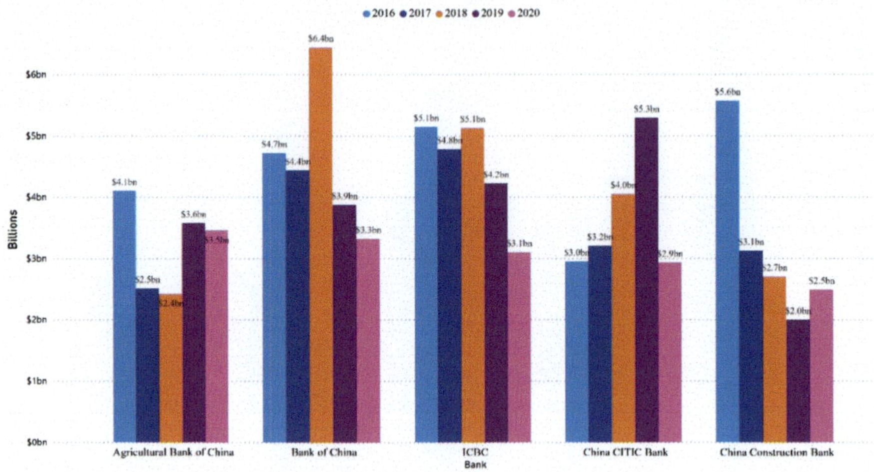

Fig. 3.3 Clustered column chart of top five banks financing the coal power sector

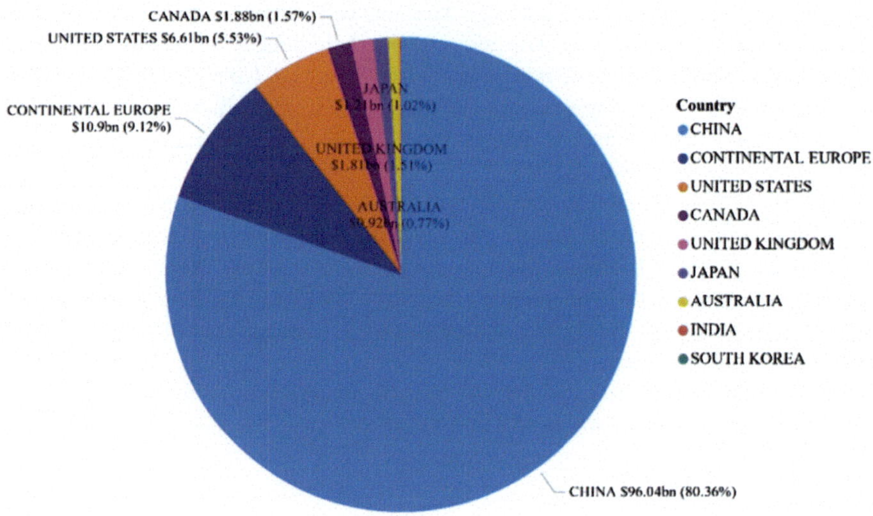

Fig. 3.4 Pie chart of investment by banks (country wise) in the coal mining sector

Currently, Canada is the major hub for the production of tar sand, as shown in Fig. 3.6. Four out of the top five banks financing the tar sands sector are of Canadian origin. Toronto-Dominion Bank (TD) leads in financing tar sands with a total investment of $24.222 billion, closely followed by Royal Bank of Canada at $22.523 billion. JP Morgan, CIBC, and Bank of Montreal are next in line, having an investment of $12.141 billion, $11.032 billion, and $10.045 billion, respectively.

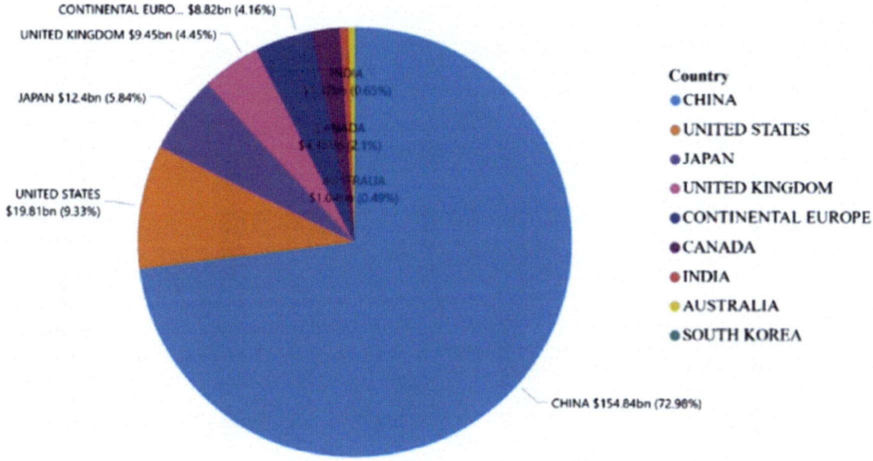

Fig. 3.5 Pie chart of investment by banks (country wise) in the coal power sector

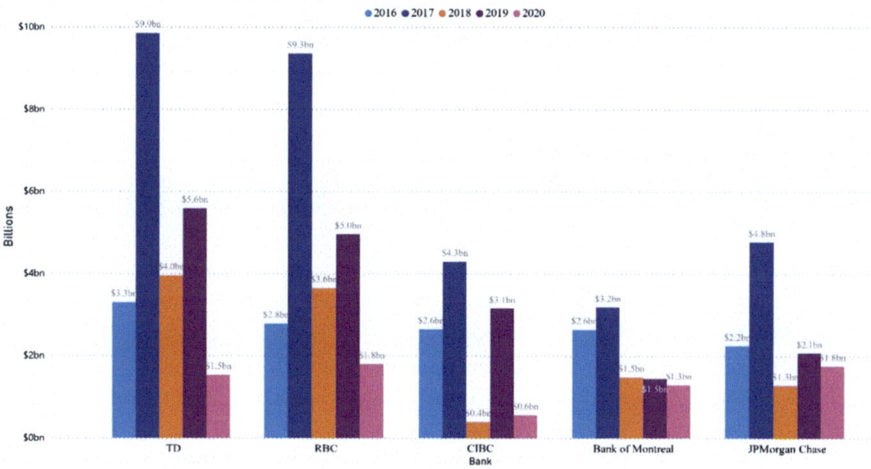

Fig. 3.6 Clustered column chart of top five banks financing tar sands

C. **Arctic Oil and Gas Sector**

The Arctic Ocean holds some of the world's largest untouched oil and natural gas reserves under its sea bed. The US Geological Survey states that about 30% of the world's undiscovered gas, 13 percent of the world's undiscovered oil, and few onshore resources are available at the north of the Arctic Circle [5]. Oil and gas production poses a significant risk to the environmental ecosystem. It also contributes to climate change through increased GHG emissions. At present, considerable oil and gas activity is being carried out in Canada, the USA, Norway, and Russia [6]. The largest oil and gas reserves in the Arctic region are found in Russia, where 43

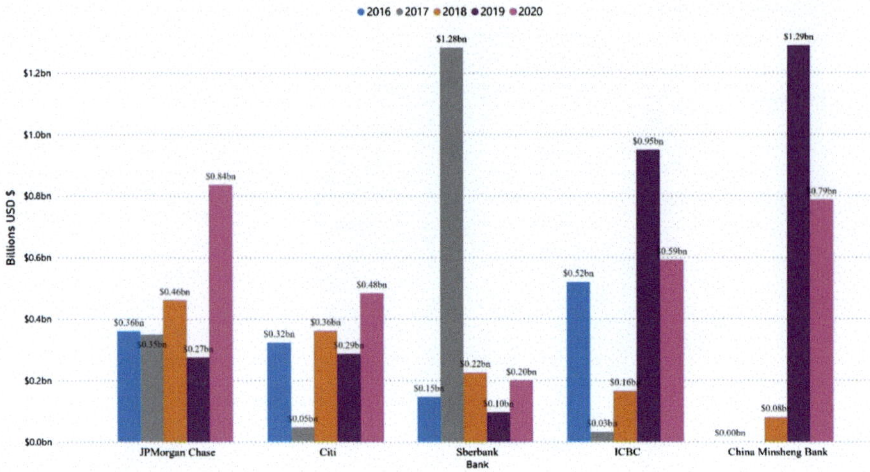

Fig. 3.7 Clustered column chart of top five banks financing arctic oil and gas sector

large oil and gas fields have been identified [6]. Figure 3.7 shows the top five banks investing in the Arctic Oil and Gas sector, indicating the increasing investment trend in this sector.

D. **Offshore Oil and Gas Sector**

Offshore oil and gas extraction refers to the development of fields and natural gas deposits under the ocean [7]. It is a critical component of the world's energy supply. Offshore production accounts for 30% of global oil production and 27% of global gas production [8]. Despite the technological advances, the process involved in oil and gas production requires huge investments. Some of the major consequences of the offshore oil and gas sector on climate change are changes in carbon dioxide levels and ocean acidity, atmospheric and water (ocean/sea) temperature, precipitation pattern and runoff, mean sea level, storm intensity, and wave regime [9]. Figure 3.8 presents the investment made by the top 05 banks in the offshore oil and gas sector. It is evident from the figure that the investment in this sector has been increasing over the years, more so in the last 2 years. BNP Paribas is leading this investment, followed by Citi bank.

E. **Fracked Oil and Gas Sector**

Fracking is the process of drilling down the earth's surface to enable the extraction of oil and natural gas from shale and other forms of rock [10]. Shale is basically a soft stratified sedimentary rock created due to the deposition of clay and mud over a period. Large quantities of water, chemical, and sand are injected into these formations at high pressure, which allows the rock to crack, helping the oil and gas to flow to the surface [10]. Fracking allows companies to access the resources of oil and gas, which are difficult to reach. It significantly boosts oil and natural gas production.

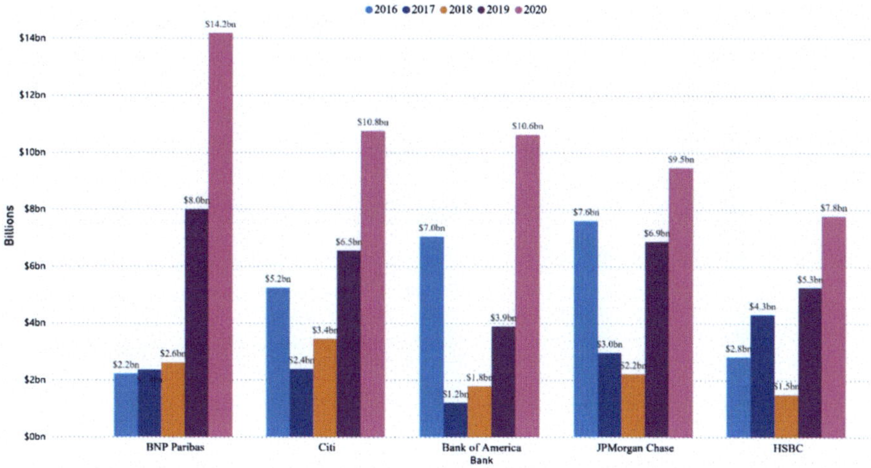

Fig. 3.8 Clustered column chart of top five banks financing offshore oil and gas sector

The effects of fracking on the environment include the release of pollutants in the air, mainly methane, impact on groundwater supply and its quality, as well as long-term effects on soil and surrounding vegetation. Fracking also has an effect on human health, forest lands, agricultural lands, grasslands, and habitats for animal and plant species near fracking sites.

In terms of investments, it is observed from Fig. 3.9 that the top four financiers of the fracking industry are all US Banks. Wells Fargo is the topmost bank financing in this sector, with a total investment of $54 billion since 2016. It is followed by JP Morgan Chase, Citi, Bank of America, and Barclays, with investments of $52.2 billion, $38.92 billion, $38.9 billion, and $23.9 billion, respectively. A positive trend is seen during 2020 compared to 2019 in terms of reduction in investments for all the top 04 leading banks financing Fracked Oil and Gas sector.

F. **Liquefied Natural Gas (LNG) Sector**

LNG is a clear, colorless, and non-toxic liquid that forms when natural gas is cooled to −162 °C (−260 °F). The cooling process shrinks the gas, and then it is transferred over to the sea, after which it is re-gasified and fed into the infrastructure of the local gas pipeline [11]. Although LNG contains less carbon per unit of energy than conventional fuel, its use might not reduce GHG emissions on a life-cycle basis [12]. The GHG emissions from the whole life-cycle of natural gas, consisting of extraction, transportation, liquefaction, and re-gasification, are very high [13]. LNG predominantly consists of methane, and leakages of only 2–3% make LNG as bad for the environment as that of coal [14]. The environmental impacts of drilling for gas, LNG carriers, terminals, and waste disposal facilities on public health are very high.

From Fig. 3.10, it is found that the top banks financing this sector include Morgan Stanley and Citi, with a total of $8.6 billion and $8.1 billion, respectively. JP Morgan

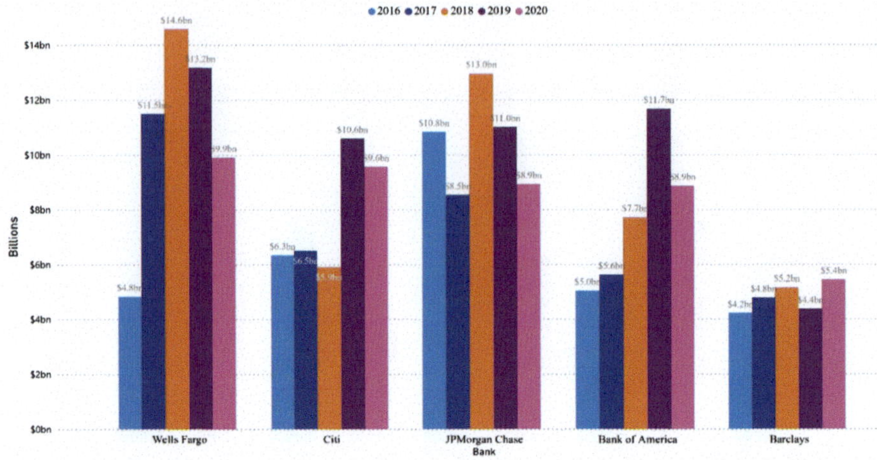

Fig. 3.9 Clustered column chart of top five banks financing fracked oil and gas sector

Chase, the Bank of America, and SMBC Group are the next top financiers of this sector with a total of $7.81 billion, $6.55 billion, and $6.54 billion, respectively. It is also observed that in the last couple of years, the investment in this sector made by JP Morgan Chase, Citi, SMBC Group, and the Bank of America has increased.

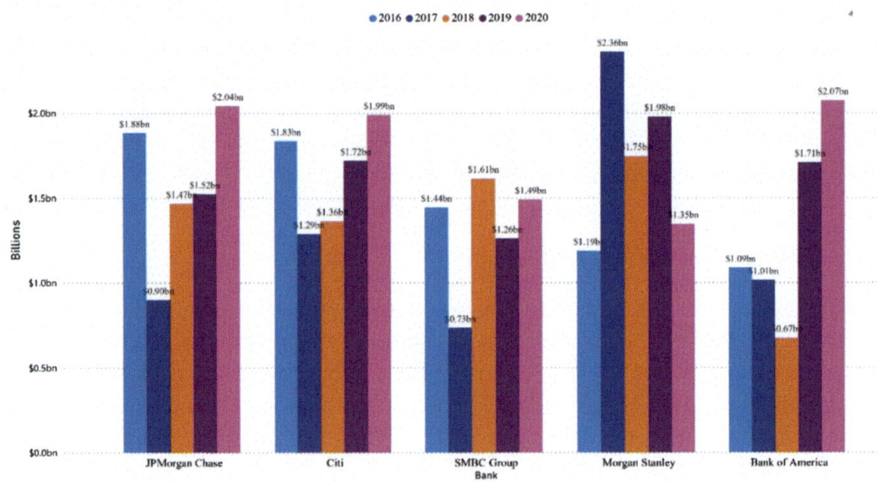

Fig. 3.10 Clustered column chart of top five banks financing the LNG sector

3.2.2 Analysis of Various Features Using Statistical Methods

In the following sub-sections, the dataset is visualized and analyzed using various statistical methods such as stacked bar graph, box plot, and correlation analysis.

A. Stacked Bar Graph and Pie Chart

A Stacked Bar Graph (or Stacked Bar Chart) consists of multiple bars series stacked horizontally one after the other. Each bar in a standard bar chart is divided into a number of sub-bars stacked end to end. The length of the bar indicates the numeric value and makes it easy to follow the variation of all the variables presented side by side. The stacked bar graph is used to derive a basic understanding of the distribution of countries and banks in general and the State Bank of India in particular. A Pie Chart is a circular statistical graph which is divided into portions, and each of them represents the percentage or count of the observations.

(i) Countries and Banks

The dataset used for this research includes top banks from different countries. Banks have been assigned to a particular country as per their origin or headquarter. For Europe, the United Kingdom has been considered separately, and the rest of all European countries have been included under "Continental Europe". Figure 3.11 shows the distribution of banks per country, with the values of individual banks financing the fossil fuel sector along with the total financing values of the country.

It needs to be noted that the dataset is limited due to the smaller number of disclosures in the Banking, Financial Services, and Insurance (BFSI) sector. It is evident from Fig. 3.11 that the banks in the United States are leading the investments

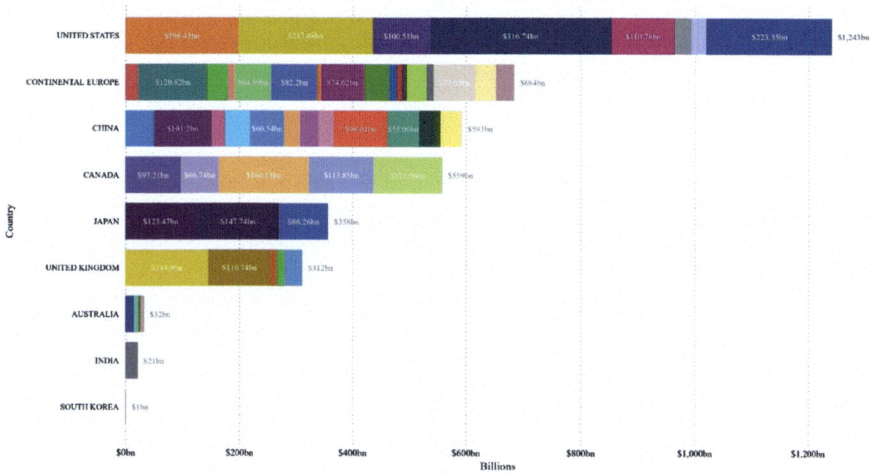

Fig. 3.11 Stacked bar graph showing country-wise investment by banks

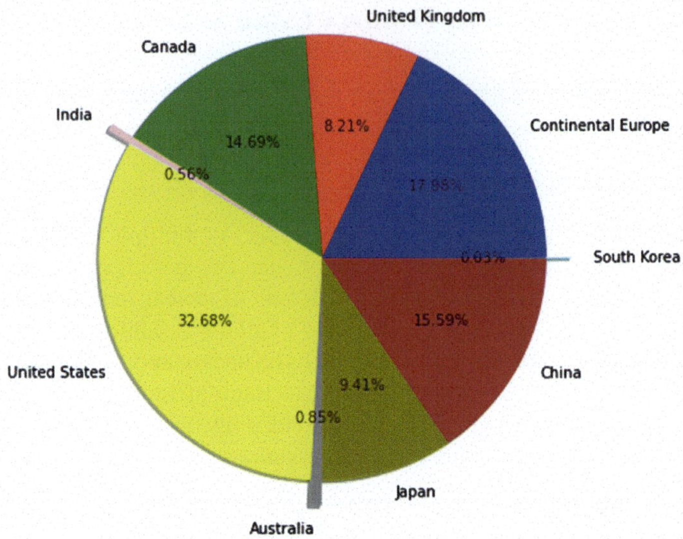

Fig. 3.12 Pie chart of banks and their share of investment

in the fossil fuel sector, followed by Continental Europe and China. Figure 3.12 presents the same fact in terms of the financial share of each country in the dataset.

(ii) **India—State Bank of India (SBI)**

For India, the data is available only for the State Banks of India. Figure 3.13 demonstrates the share of investment made by SBI in different companies in the fossil fuel sector. It is clear from the pie chart below that a significant investment has been made in the oil and gas sector. Oil and Natural Gas Corporation Limited gets the maximum financial support (21.37%), followed by Bharat Petroleum Corporation Limited (14.98%) and Hindustan Petroleum Corporation Limited (13.52%).

B. **Box Plot**

Box plots visually show the distribution of numerical data and skewness by displaying the data quartiles (or percentiles) and averages. They are used to show overall patterns of response for a group. They provide a useful way to visualize the range and other characteristics of responses for a large group [15]. Figure 3.14 presents the distributional aspects of a group of scores as well as the level of the scores, which is further divided into four equal-sized groups (25%) made from the ordered scores. The lines dividing the groups are called quartiles, and the groups are referred to as quartile groups, which are labeled as 1–4, starting at the bottom. Hence the five terms can be defined below:

1. Median—The median (middle quartile) marks the mid-point of the data and is shown by the line that divides the box into two parts. Half the scores are greater than or equal to this value, and half are less.

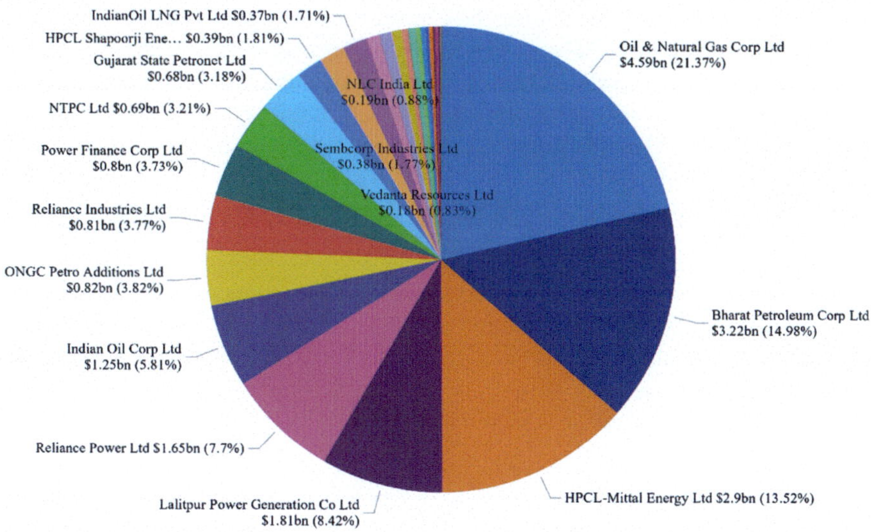

Fig. 3.13 Pie chart of investment by State Bank of India in various companies

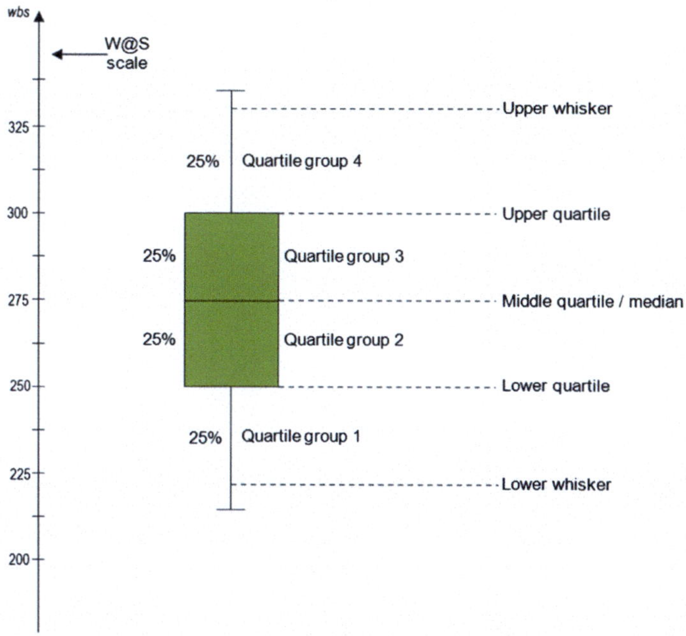

Fig. 3.14 Box plot with four equal-sized groups naming upper whisker-upper quartile, upper quartile-median, median-lower quartile, lower quartile-lower whisker

2. Inter-quartile range—The middle "box" represents the group's middle 50 percent of scores. The range of scores from lower to upper quartile is referred to as the inter-quartile range. The middle 50% of scores fall within the inter-quartile range.
3. Upper quartile—Seventy-five percent of the scores fall below the upper quartile.
4. Lower quartile—Twenty-five percent of scores fall below the lower quartile.
5. Whiskers—The upper and lower whiskers represent scores outside the middle 50 percent. Whiskers often (but not always) stretch over a wider range of scores than the middle quartile groups.

The individual box plot for the total fossil fuel investments of 60 banks in the 7 sectors mentioned above has been plotted and presented in Fig. 3.15. The libraries like Pandas, NumPy, Matplotlib, and Seaborn are imported to conduct the exploratory data analysis. The code snippet to draw the box plot and to get the results using different colors for each sector is shared below:

```
import pandas as pd
import numpy as np
import seaborn as sns
import matplotlib.pyplot as plt
import matplotlib.mlab as mlab
import matplotlib
plt.style.use('ggplot')
from matplotlib.pyplot import figure
df = pd.read_excel (r"C:\Users\Acer\Downloads\Climate
Change\Dataset\Total and Sector Wise Fossil Fuel Fianace.xlsx")
df
fig, axes = plt.subplots(4, 2, figsize=(13, 15), sharey=True)
sns.boxplot(x='Total Fossil Fuel', data=df, ax= axes[0,0],
color='blue')
sns.boxplot(x='Tar Sands', data=df, ax= axes[0,1], color='orange')
sns.boxplot(x='Arctic Oil & Gas', data=df, ax= axes[1,0],
color='green')
sns.boxplot(x='OffShore Oil & Gas', data=df, ax= axes[1,1],
color='brown')
sns.boxplot(x='Fracked Oil & Gas', data=df, ax= axes[2,0],
color='purple')
sns.boxplot(x='LNG', data=df, ax= axes[2,1], color='yellow')
sns.boxplot(x='Coal Mining', data=df, ax= axes[3,0], color='olive')
sns.boxplot(x='Coal Power', data=df, ax= axes[3,1], color='red')
```

Some of the observations from the above box plots in Fig. 3.15 are mentioned below:

- Tar Sands and Coal Mining sectors have the maximum number of outliers, followed by Coal Power and Fracked Oil and Gas sectors. This clearly indicates the presence of noise in the dataset.
- The range of outliers observed in each sector is as follows:
  - Total Fossil Fuel—beyond \$0.2 trillion.
  - Offshore Oil & Gas—beyond \$20 billion.

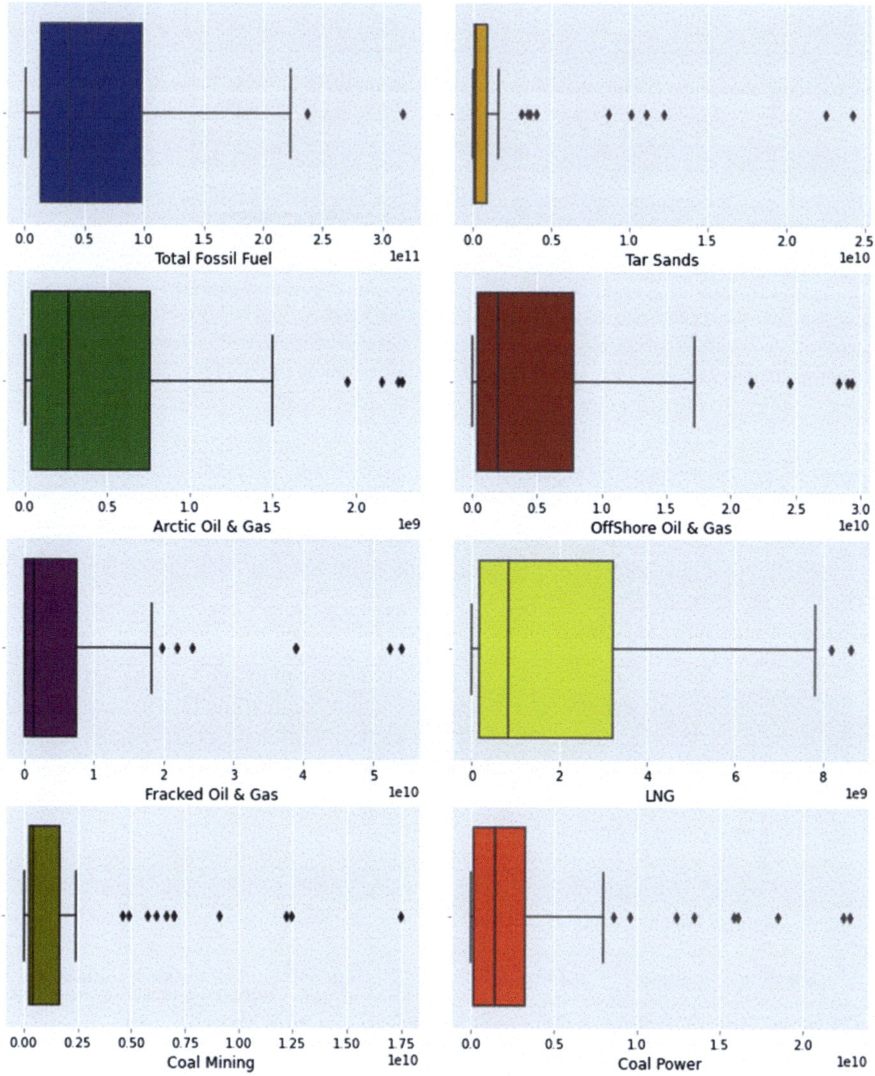

Fig. 3.15 Box plot for total investment of 60 banks in fossil fuel sectors from the year 2016–2020

- Fracked Oil & Gas—above \$15 billion.
- Tar Sands—above \$10 billion.
- LNG—beyond \$7 billion.
- Coal Power—beyond \$5 billion.
- Coal Mining—beyond \$2.5 billion.
- Arctic Oil and Gas—beyond \$1.5 billion.

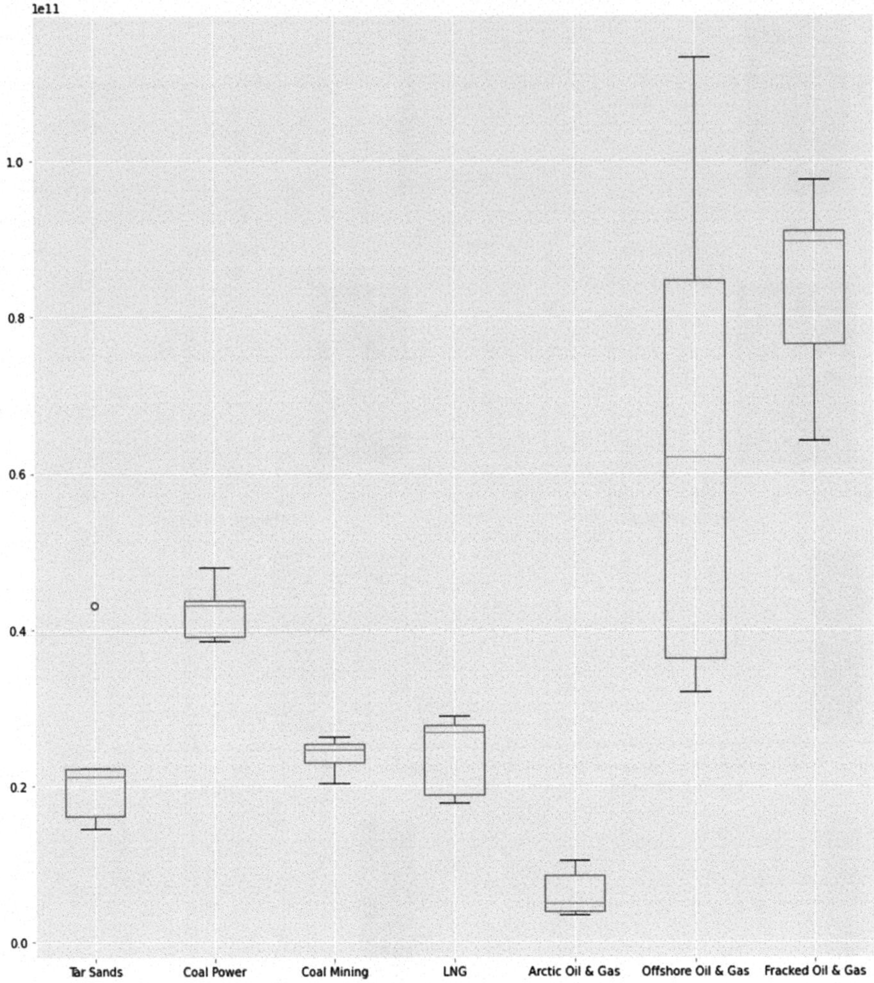

Fig. 3.16 Box plot of grouped sector of fossil fuel for the year 2016–2020

- LNG, Offshore Oil and Gas, and Arctic Oil and Gas sectors are in the Q3 range from median to upper quartile.
- Total Fossil fuel is also in the Q3 range from median to upper quartile.

After presenting sector-wise box plot, a holistic view is demonstrated in Fig. 3.16 through the box plot for grouped fossil fuel sectors. The box plot is created using the below code:

```
import numpy as np
import seaborn as sns
```

```
import matplotlib.pyplot as plt
import matplotlib.mlab as mlab
import matplotlib
plt.style.use('ggplot')
from matplotlib.pyplot import figure
df = pd.read_excel (r"C:\Users\Acer\Downloads\Climate Change\Banks & Climate Change\Dataset\climate_change_main_dataset.xlsx", sheet_name='Sheet20')

fig, axes = plt.subplots(1, 1, figsize=(13, 15), sharey=True)
df.boxplot(column=['Tar Sands','Coal Power','Coal Mining','LNG','Arctic Oil & Gas','Offshore Oil & Gas','Fracked Oil & Gas'])
```

In Fig. 3.16, the x-axis represents seven sectors and the y-axis represents the investment by these sectors in US million dollars. It clearly indicates that the investment made by Offshore Oil and Gas sector is the maximum, and Fracked Oil and Gas sector is the highest.

3.2.3 Correlation Analysis

In this section, relations between different variables are evaluated with the help of Scatter Plots, Heat Maps, and Regression Plots.

A. **Scatter Plot**

A scatter plot uses dots to represent values for two different numeric variables. The position of each dot on the horizontal and vertical axis indicates values for an individual data point. Scatter plots are used to observe relationships between variables [16]. The investment made by banks in the fossil fuel sectors is leading towards the biggest hazard to climate change. The scatter plot is used to analyze total fossil fuel investment and also sector-wise investment by 60 banks in the database. The below Python code is written to create scatter plots:

```
import matplotlib.pyplot as plt
import pandas as pd

fig, ax = plt.subplots(4,2, figsize=(13, 17))
ax[0,0].scatter(x = df['Tar Sands'], y = df['Total Fossil Fuel'])
ax[0,0].set_xlabel("Tar Sands")
ax[0,0].set_ylabel("Total Fossil Fuel")

ax[0,1].scatter(x = df['Arctic Oil & Gas'], y = df['Total Fossil Fuel'])
ax[0,1].set_xlabel("Arctic Oil & Gas")
ax[0,1].set_ylabel("Total Fossil Fuel")

ax[1,0].scatter(x = df['OffShore Oil & Gas'], y = df['Total Fossil Fuel'])
```

```
ax[1,0].set_xlabel("OffShore Oil & Gas")
ax[1,0].set_ylabel("Total Fossil Fuel")

ax[1,1].scatter(x = df['Fracked Oil & Gas'], y = df['Total Fossil
Fuel'])
ax[1,1].set_xlabel("Fracked Oil & Gas")
ax[1,1].set_ylabel("Total Fossil Fuel")

ax[2,0].scatter(x = df['LNG'], y = df['Total Fossil Fuel'])
ax[2,0].set_xlabel("LNG")
ax[2,0].set_ylabel("Total Fossil Fuel")

ax[2,1].scatter(x = df['Coal Mining'], y = df['Total Fossil Fuel'])
ax[2,1].set_xlabel("Coal Mining")
ax[2,1].set_ylabel("Total Fossil Fuel")

ax[3,0].scatter(x = df['Coal Power'], y = df['Total Fossil Fuel'])
ax[3,0].set_xlabel("Coal Power")
ax[3,0].set_ylabel("Total Fossil Fuel")

plt.show()
```

Figure 3.17 depicts the relationship between total fossil fuel investment and sector-wise investment during 2016–2020, with the help of scatter plots. The x-axis represents the investment of the respective sector and the y-axis represents the total fossil fuel investment.

Figure 3.17 shows that there is a significant relationship between investment in each sector and the total investment in fossil fuels. All sectors show that there is a positive relation with total investment, majorly showing two patterns of relationship. The first pattern is observed in the "LNG", "Fracked Oil and Gas", and "Offshore Oil and Gas" sectors, and the second pattern is seen in the "Coal Power", and "Coal Mining" sectors.

B. **Heat Map Correlation Matrix**

The correlation between the top 60 Banks with their total fossil fuel and sector-wise investment has been plotted in the Heat Map Correlation Matrix. A correlation matrix is simply a table that displays the correlation coefficients for different variables. The matrix depicts the correlation between all the possible pairs of values in a table. It is a powerful tool to summarize a large dataset and identify and visualize patterns in the given data. A correlation matrix consists of rows and columns that show the variables. Each cell in a table contains the correlation coefficient [17]. A correlation heatmap is a graphical representation of a correlation matrix representing the relation between different variables.

The heat map correlation matrix has been plotted for investment in the fossil fuel sector and its consequences for each year from 2016 to 2020 by using Pearson correlation which is a measure of linear correlation between two variables. It has a numerical value between -1.0 and $+1.0$. Figure 3.18 presents the distribution of each feature and the relationship between them. Below is the snippet of Python code used for generating the Heat Map.

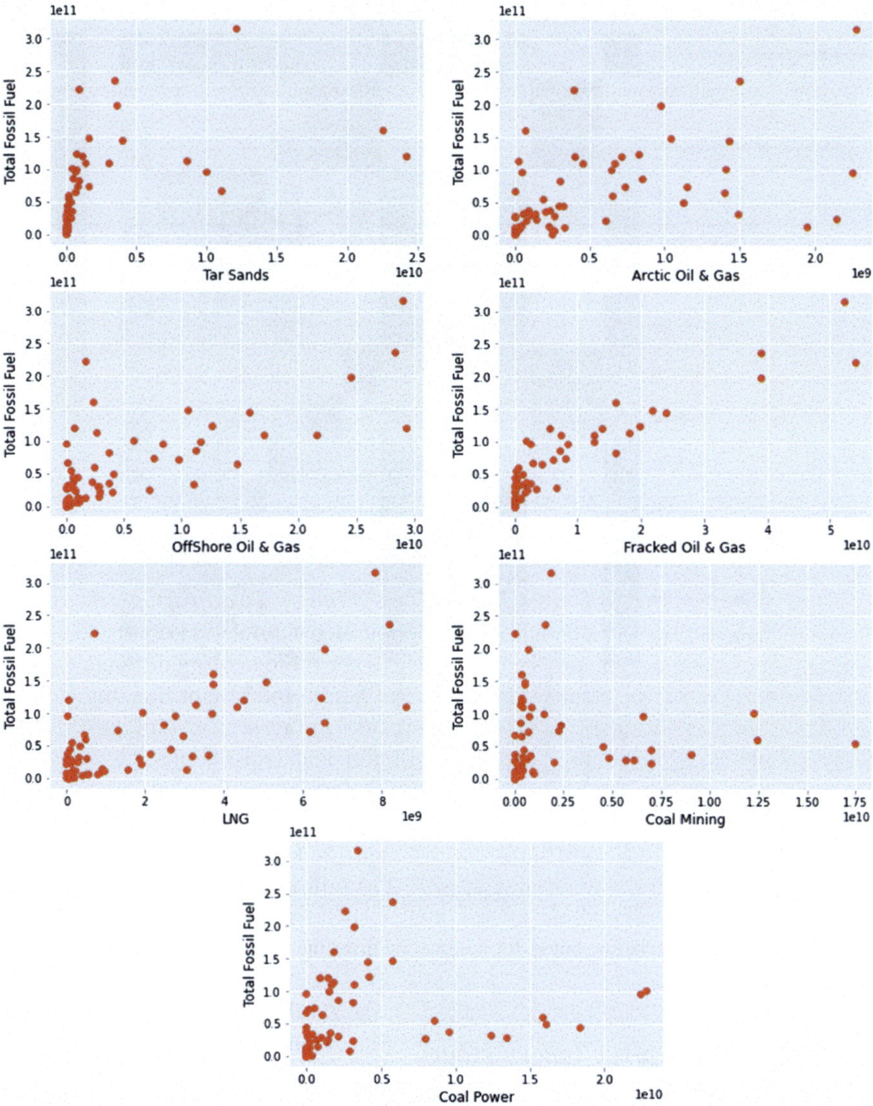

Fig. 3.17 Scatter plot for a total investment of each sector compared to total fossil fuel investment for 60 banks

```
import pandas as pd
import numpy as np
import seaborn as sns

import matplotlib.pyplot as plt
import matplotlib.mlab as mlab
import matplotlib
plt.style.use('ggplot')
```

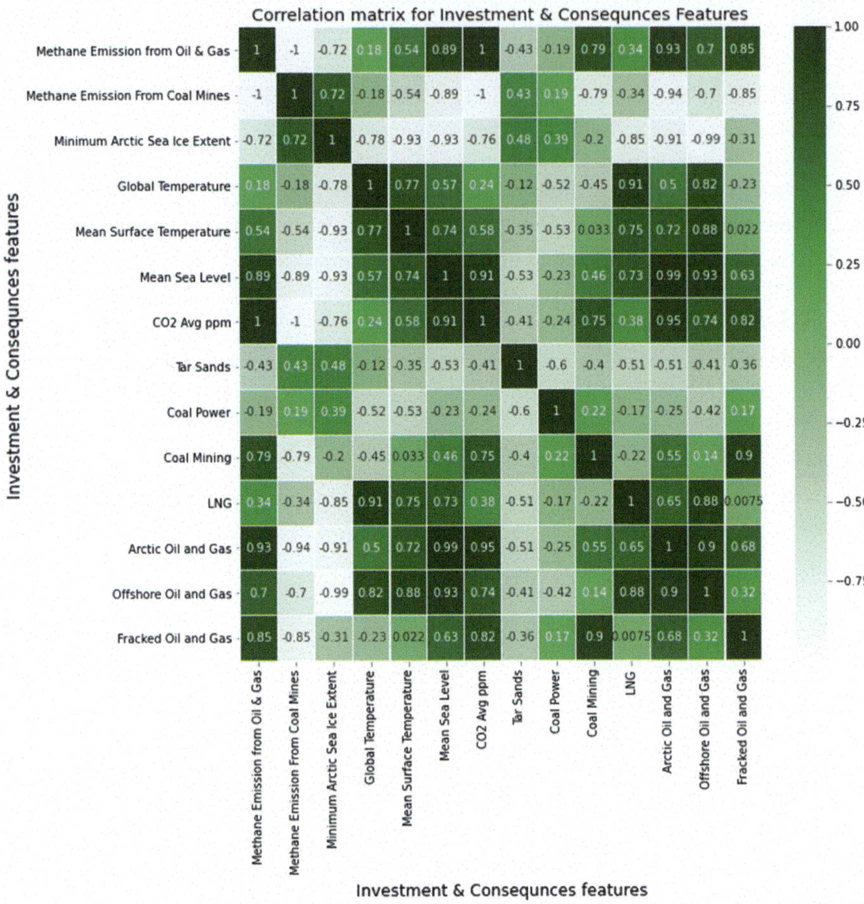

Fig. 3.18 Heat map correlation matrix for sector-wise investment and the consequences (2016–2020)

```
from matplotlib.pyplot import figure

df = pd.read_excel(r"C:\Users\Acer\Downloads\Climate Change\Banks &
Climate          Change\Dataset\climate_change_main_dataset.xlsx",
sheet_name='Sheet7')
df

df.corr(method ='pearson')

correlation_matrix = df.corr(method ='pearson')
fig, ax = plt.subplots(figsize=(10,10))
sns.heatmap(correlation_matrix, annot = True,linewidths=0.1, ax=ax
, cmap="Greens")

plt.title("Correlation  matrix  for  Investment  &  Consequnces
Features")
```

```
plt.xlabel("Investment & Consequnces features")
plt.ylabel("Investment & Consequnces features")
plt.show()
```

Figure 3.18 shows the Pearson correlation between the sector-wise investment and its consequences. The following insights are derived from Fig. 3.18:

(1) It is observed that CO_2 Avg PPM is positively correlated with Offshore Oil and Gas (0.74), Coal Mining (0.75), Fracked Oil and Gas (0.82), and Arctic Oil and Gas (0.95). This implies that the increase in investment in these four sectors will lead to a higher CO_2 avg PPM level.
(2) Mean sea level is positively correlated with Fracked Oil and Gas (0.63), LNG (0.73), Offshore Oil and Gas (0.93), and Arctic Oil and Gas (0.99). This clearly indicates that if investment in these four sectors goes up, then it will result in an increase in the mean sea level.
(3) Mean surface temperature is positively correlated with Arctic Oil and Gas (0.72), LNG (0.75), and Offshore Oil and Gas (0.88). Mean surface temperature will increase with higher investment in these three sectors.
(4) Global temperature is positively correlated with Offshore Oil and Gas (0.82) and LNG (0.91). So, investment in these two sectors will have a substantial effect on global temperature.
(5) The minimum Arctic Sea ice extent is the sea ice extent measured at the end of the summer melting season, generally in September. It is negatively correlated with Offshore Oil and Gas (−0.99), Arctic Oil and Gas (−0.91), and LNG (−0.85). So, the greater the investments in fossil fuel sectors, the lesser the Arctic Sea ice extent.
(6) Methane emission from coal mines is positively correlated with Coal Power (0.19) and Tar Sands (0.43). Although there is a positive correlation between Coal Power and Tar Sands, the impact of investment on methane emission is weak.
(7) Methane emission from Oil & Gas is positively correlated with Offshore Oil and Gas (0.7), Coal Mining (0.79), Fracked Oil and Gas (0.85), and Arctic Oil and Gas (0.93). So, the investment in these four sectors will have higher methane emissions.

C. **Regression Plot (Reg Plot)**

The regression plot in seaborn (data visualization library) is primarily intended to add a visual guide that helps to emphasize patterns in a dataset during exploratory data analysis. A regression plot creates a regression line between two parameters and helps to visualize their linear relationships. This method is used to plot the investment in seven sectors by the top 60 banks during 2016–2020 and to generate a linear regression model fit to analyze the variation. The reg plot for total fossil fuel investment with respect to investment by each sector can be seen in Fig. 3.19. The following Python code snippet is applied for creating the Reg Plot.

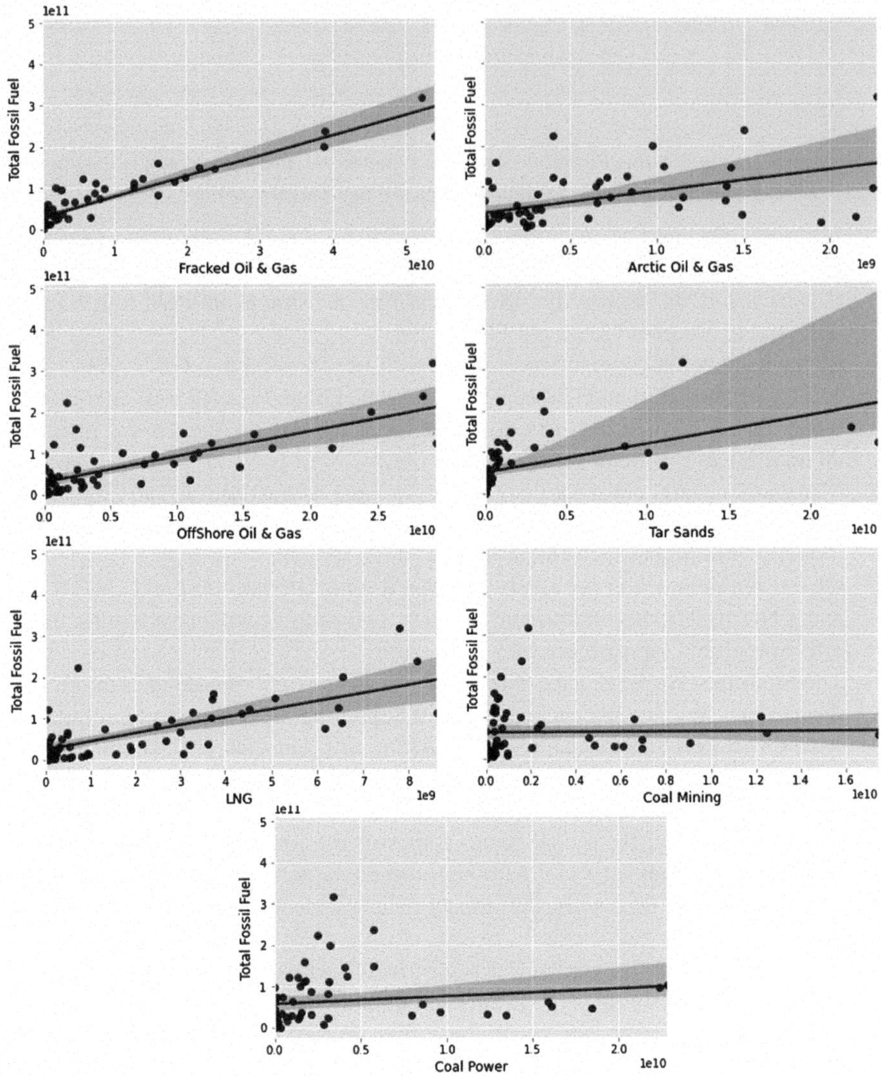

Fig. 3.19 Reg plot for total fossil fuel investment versus sector-wise investment of top 60 banks from 2016–2020

```
import pandas as pd
import numpy as np
import seaborn as sns

import matplotlib.pyplot as plt
import matplotlib.mlab as mlab
import matplotlib
plt.style.use('ggplot')
from matplotlib.pyplot import figure
```

```
df = pd.read_excel(r'C:\Users\Acer\Downloads\Climate Change\Banks & Climate Change\Dataset\Total and Sector Wise Fossil Fuel Fianace.xlsx')
df

fig, axes = plt.subplots(4, 2, figsize=(14, 17), sharey=True)

sns.regplot(x="Fracked Oil & Gas", y="Total Fossil Fuel", data=df,color="black", ax = axes[0,0])
sns.regplot(x="Arctic Oil & Gas", y="Total Fossil Fuel", data=df, color="black",ax = axes[0,1])
sns.regplot(x="OffShore Oil & Gas", y="Total Fossil Fuel", data=df, color="black",ax =axes[1,0])
sns.regplot(x="Tar Sands", y="Total Fossil Fuel", data=df, color="black",ax = axes[1,1])
sns.regplot(x="LNG", y="Total Fossil Fuel", data=df, color="black",ax = axes[2,0])
sns.regplot(x="Coal Mining", y="Total Fossil Fuel", data=df,color="black", ax = axes[2,1])
sns.regplot(x="Coal Power", y="Total Fossil Fuel", data=df, color="black",ax = axes[3,0])
```

From Fig. 3.19, it is seen that investments in total fossil fuel vary less with respect to the change in investments in coal power and coal mining sectors. In contrast, a greater share in the investment of total fossil fuel is observed in Tar Sands, Fracked Oil and Gas, Arctic Oil and Gas, Offshore Oil and Gas, and LNG.

3.2.4 Summary of Exploratory Data Analysis

In the first section of EDA, all fossil fuel sectors, with respect to banks, have been plotted. Different features are analyzed in Sect. 3.2 of EDA to study their trends and impacts. Section 3.3 deals with correlation analysis and feature selection. Now the dataset is processed, prepared, and ready for model building and evaluation.

3.3 Model Building

Due to investment in fossil fuels like Tar Sands, Coal Mining, Coal Power, and so on, CO_2 emission has been increasing continuously. Similarly, due to investment in Fracked Oil and Gas, Offshore Oil and Gas, and Arctic Oil and Gas, methane emissions have also been rising rapidly. These investments are impacting mean surface temperature, global temperature, Arctic Sea ice extent, and mean sea level adversely, which clearly indicates that all these consequences are interconnected. In this section, linear regression with multiple variables, also known as the multivariate regression model, is built to predict these consequences on the basis of future investment in these sectors. In order to predict these consequences, two steps are involved, which are mentioned as follows:

1. The first step is to build models using the current dataset and train them to predict future investment.
2. The second step is to use the predicted future investment derived from step 1 to predict the future consequences.

3.3.1 Multivariate Regression Analysis and Model Fitting

Multivariate regression is used to predict an output using multiple independent variables. The ability of a machine learning model to generalize data comparable to that with which it was trained is measured by model fitting.

Regression Line or Best-Fit Line

The first step is to predict the investment by these banks in each sector for the year 2021 using the current dataset and applying regression or best-fit line. In order to achieve this, the following equations are used:

$$b = \overline{y} - m\overline{x} \tag{3.1}$$

where m = best-fit slope and b = best-fit y-intercept

$$\text{Regression Line} = [(m * x) + b \text{ for } x \text{ in } xs] \tag{3.2}$$

where xs = input sample on the x-axis

The process of model building starts by importing the NumPy library and the "mean" function from the statistics library. The next step is to input the datasets ys and xs as sector investment and year, respectively. A function has been defined for best-fit slope and intercept, followed by the calculation of the values. Then using Eq. (3.2) and scatter plot, the points are plotted, and the regression line is drawn to get the visualization. Next, after getting the plot, the output is predicted using the year value from the Eq. (3.1).

The second step is to predict investment by banks for the year 2022. While predicting investments for 2021, there are only five data points from the year 2016 to 2020. Due to the scarcity of data, estimated investments for the year 2021 have been added to the original dataset, and now this dataset is used to predict investments for the year 2022. Similarly, sector-wise investments for the year 2023 have been predicted. The following sub-section has given the Python code and best-fit line diagram representing the sector-wise investment for the year 2023.

i. **Regression Line for Tar Sands Investment Prediction:**

The Python code for the regression line for Tar Sands investment is provided below:

```
from statistics import mean
import numpy as np

tar_list = df["Tar Sands"].tolist()
year_list = df["Year"].tolist()

ys = np.array(tar_list, dtype=np.float64)
xs = np.array(year_list, dtype=np.float64)

def best_fit_slope_and_intercept(xs,ys):
m = (((mean(xs)*mean(ys)) - mean(xs*ys)) /
((mean(xs)*mean(xs)) - mean(xs*xs)))

b = mean(ys) - m*mean(xs)

return m,b

m, b = best_fit_slope_and_intercept(xs,ys)

print(m,b)
-3079182467.470703 6237217165803.536

regression_line = [(m*x)+b for x in xs]

import matplotlib.pyplot as plt
from matplotlib import style
style.use('ggplot')

plt.scatter(xs,ys,color='black', label='data')
plt.plot(xs, regression_line, color='red', label='regression line')
plt.legend(loc=1)
plt.title('Best-Fit Line for Tar Sands')
plt.xlabel('Year')
plt.ylabel('US Billion Dollar')
plt.show()
```

The below code is to predict the Total Tar Sands investment by various banks for the year 2023:

```
predict_x1 = 2023
predict_y1 = (m*predict_x1)+b
print(predict_y1)
8031034110.303711
```

After computing, the values of y-intercept and slope are b = 6,237,217,165,803.536 and m = −3,079,182,467.470703. A regression line has been plotted as shown in Fig. 3.20 and investment for the year 2023 is $8.03 billion.

ii. **Regression line for coal power investment prediction:**

The Python code for the regression line for Coal Power investment is provided below:

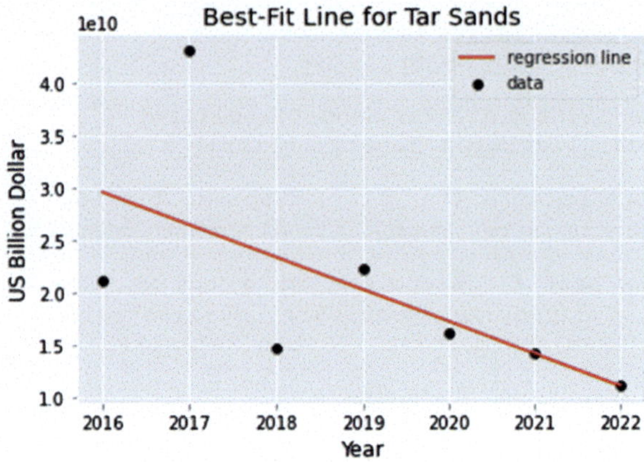

Fig. 3.20 Regression line for total tar sands investment by banks from the year 2016 to 2022

```
from statistics import mean
import numpy as np

coal_power_list = df["Coal Power"].tolist()
year_list = df["Year"].tolist()

ys = np.array(coal_power_list, dtype=np.float64)
xs = np.array(year_list, dtype=np.float64)

def best_fit_slope_and_intercept(xs,ys):
m = (((mean(xs)*mean(ys)) - mean(xs*ys)) /
((mean(xs)*mean(xs)) - mean(xs*xs)))

b = mean(ys) - m*mean(xs)

return m,b

m, b = best_fit_slope_and_intercept(xs,ys)

print(m,b)
-455566070.0546875 961767175019.3501

regression_line = [(m*x)+b for x in xs]

import matplotlib.pyplot as plt
from matplotlib import style
style.use('ggplot')

plt.scatter(xs,ys,color='black', label='data')
plt.plot(xs,  regression_line,  color='blue',  label='regression
line')
plt.legend(loc=1)
plt.title('Best-Fit Line for Coal Power')
plt.xlabel('Year')
plt.ylabel('US Billion Dollar')
plt.show()
```

The below code is to predict the Coal Power investment by various banks for the year 2023:

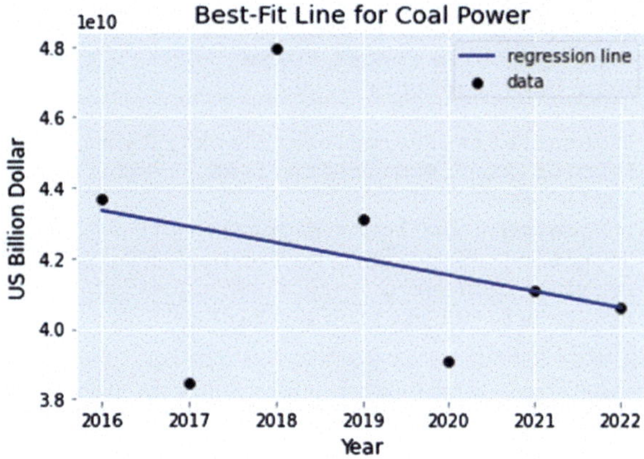

Fig. 3.21 Regression line for total coal power investment by the banks from the year 2016 to 2022

```
predict_x1 = 2023
predict_y1 = (m*predict_x1)+b
print(predict_y1)
40157015298.717285
```

After computing, the values of y-intercept and slope are b = 961,767,175,019.3501 and m = −455,566,070.0546875. A regression line has been plotted as shown in Fig. 3.21 and investment for the year 2023 is $40.15 billion.

iii. **Regression Line for Coal Mining Investment Prediction:**

The Python code for the regression line for Coal Mining investment is provided below:

```
from statistics import mean
import numpy as np

coal_mining_list = df["Coal Mining"].tolist()
year_list = df["Year"].tolist()

ys = np.array(coal_mining_list, dtype=np.float64)
xs = np.array(year_list, dtype=np.float64)

def best_fit_slope_and_intercept(xs,ys):
m = (((mean(xs)*mean(ys)) - mean(xs*ys)) /
((mean(xs)*mean(xs)) - mean(xs*xs)))

b = mean(ys) - m*mean(xs)

return m,b

m, b = best_fit_slope_and_intercept(xs,ys)

print(m,b)
1162572711.8613281 -2322169344230.023

regression_line = [(m*x)+b for x in xs]
```

```
import matplotlib.pyplot as plt
from matplotlib import style
style.use('ggplot')

plt.scatter(xs,ys,color='black', label='data')
plt.plot(xs,  regression_line,  color='green',  label='regression
line')
plt.legend(loc="lower right")
plt.title('Best-Fit Line for Coal Mining')
plt.xlabel('Year')
plt.ylabel('US Billion Dollar')
plt.show()
```

The below code is to predict the Coal Mining investment by various banks for the year 2023:

```
predict_x1 = 2023
predict_y1 = (m*predict_x1)+b
print(predict_y1)
29715251865.443848
```

After computing, the values of y-intercept and slope are b = − 2, 322,169,344,230.023 and m = 1,162,572,711.8613281. A regression line has been plotted as shown in Fig. 3.22 and investment for the year 2023 is $29.71 billion.

iv. **Regression Line for LNG (Liquefied Natural Gas) Investment Prediction:**

The Python code for the regression line for LNG investment is provided below:

```
from statistics import mean
import numpy as np

lng_list = df["LNG"].tolist()
year_list = df["Year"].tolist()

ys = np.array(lng_list, dtype=np.float64)
```

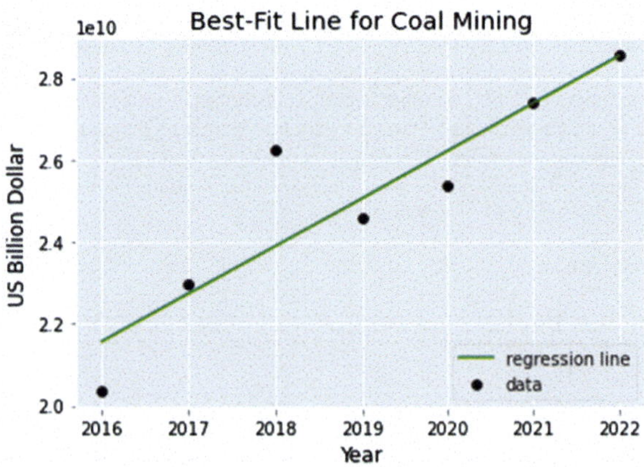

Fig. 3.22 Regression line for total coal mining investment by banks from the year 2016 to 2022

```
xs = np.array(year_list, dtype=np.float64)

def best_fit_slope_and_intercept(xs,ys):
m = (((mean(xs)*mean(ys)) - mean(xs*ys)) /
((mean(xs)*mean(xs)) - mean(xs*xs)))

b = mean(ys) - m*mean(xs)

return m,b

m, b = best_fit_slope_and_intercept(xs,ys)

print(m,b)
1123958861.296875 -2244140246161.944

regression_line = [(m*x)+b for x in xs]

import matplotlib.pyplot as plt
from matplotlib import style
style.use('ggplot')

plt.scatter(xs,ys,color='black', label='data')
plt.plot(xs, regression_line, color='purple', label='regression
line')
plt.legend(loc="lower right")
plt.title('Best-Fit Line for LNG')
plt.xlabel('Year')
plt.ylabel('US Billion Dollar')
plt.show()
```

The below code is to predict the LNG investment by various banks for the year 2023:

```
predict_x1 = 2023
predict_y1 = (m*predict_x1)+b
print(predict_y1)
29628530241.634277
```

After computing, the values of y-intercept and slope are b = −2, 244,140,246,161.944 and m = 1,123,958,861.296875. A regression line has been plotted as shown in Fig. 3.23 and investment for the year 2023 is $29.62 billion.

v. **Regression Line for Arctic Oil and Gas Investment Prediction:**

The Python code for the regression line for Arctic Oil and Gas investment is provided below:

```
from statistics import mean
import numpy as np

arctic_list = df["Arctic Oil & Gas"].tolist()
year_list = df["Year"].tolist()

ys = np.array(arctic_list, dtype=np.float64)
xs = np.array(year_list, dtype=np.float64)
```

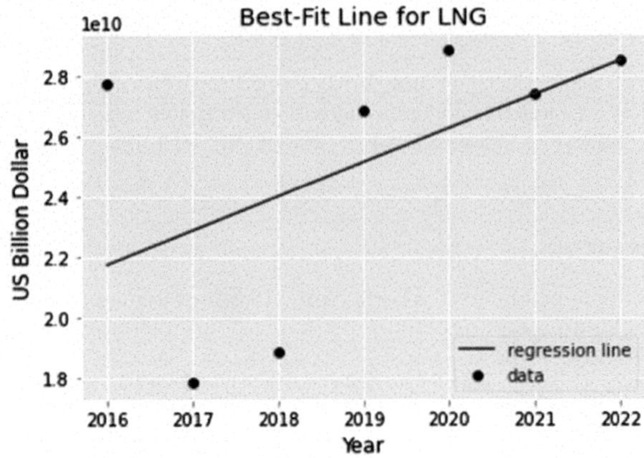

Fig. 3.23 Regression line for total LNG investment by banks from the year 2016 to 2022

```
def best_fit_slope_and_intercept(xs,ys):
m = (((mean(xs)*mean(ys)) - mean(xs*ys)) /
((mean(xs)*mean(xs)) - mean(xs*xs)))

b = mean(ys) - m*mean(xs)

return m,b

m, b = best_fit_slope_and_intercept(xs,ys)

print(m,b)
1798865319.894043 -3623870345213.084

regression_line = [(m*x)+b for x in xs]

import matplotlib.pyplot as plt
from matplotlib import style
style.use('ggplot')
plt.scatter(xs,ys,color='black', label='data')

plt.plot(xs, regression_line, color='olive', label='regression
line')
plt.legend(loc="lower right")
plt.title('Best-Fit Line for Arctic Oil & Gas')
plt.xlabel('Year')
plt.ylabel('US Billion Dollar')
plt.show()
```

The below code is to predict the Arctic Oil and Gas investment by various banks for the year 2023:

```
predict_x1 = 2023
predict_y1 = (m*predict_x1)+b
print(predict_y1)
15,234,196,932.564941
```

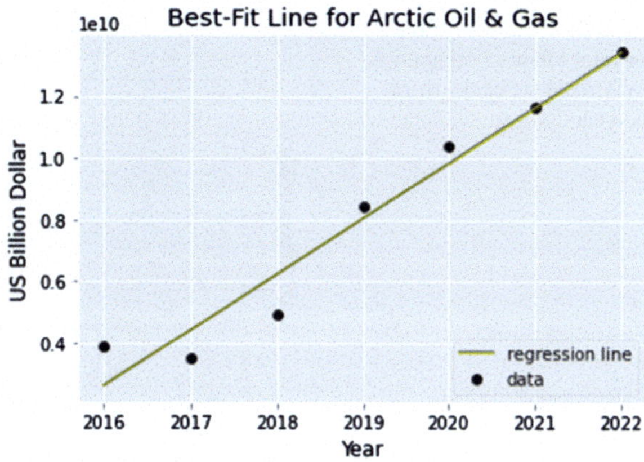

Fig. 3.24 Regression line for total arctic oil and gas investment by banks from the year 2016 to 2022

After computing, the values of y-intercept and slope are b = −3, 623,870,345,213.084 and m = 1,798,865,319.894043. A regression line has been plotted as shown in Fig. 3.24, and the investment for the year 2023 is $15.23 billion.

vi. **Regression Line for Offshore Oil and Gas Investment Prediction:**

The Python code for the regression line for Offshore Oil and Gas investment is provided below:

```
from statistics import mean
import numpy as np

offshore_list = df["Offshore Oil & Gas"].tolist()
year_list = df["Year"].tolist()

ys = np.array(offshore_list, dtype=np.float64)
xs = np.array(year_list, dtype=np.float64)

def best_fit_slope(xs,ys):
m = (((mean(xs)*mean(ys)) - mean(xs*ys)) /
((mean(xs)*mean(xs)) - mean(xs*xs)))

b = mean(ys) - m*mean(xs)

return m,b

m, b = best_fit_slope_and_intercept(xs,ys)

print(m,b)
15057249898.695312 -30319966899308.016

regression_line = [(m*x)+b for x in xs]

import matplotlib.pyplot as plt
from matplotlib import style
style.use('ggplot')

plt.scatter(xs,ys,color='black', label='data')
```

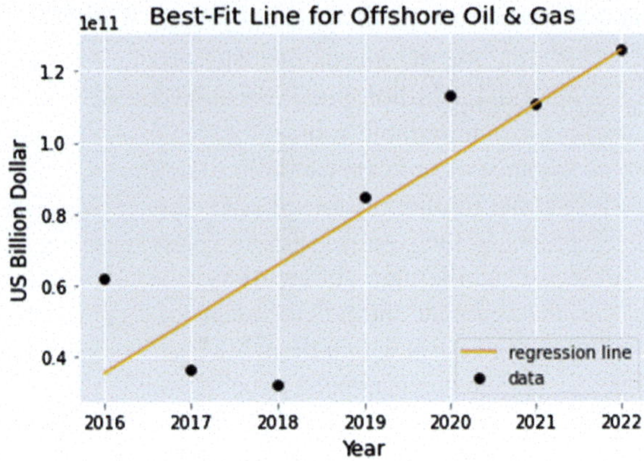

Fig. 3.25 Regression line for total offshore oil and gas investment by banks from the year 2016 to 2022

```
plt.plot(xs, regression_line, color='orange', label='regression
line')
plt.legend(loc="lower right")
plt.title('Best-Fit Line for Offshore Oil & Gas')
plt.xlabel('Year')
plt.ylabel('US Billion Dollar')
plt.show()
```

The below code is to predict the Offshore Oil and Gas investment by various banks for the year 2023:

```
predict_x1 = 2023
predict_y1 = (m*predict_x1)+b
prin(predict_y1)
140849645752.60156
```

After computing, the values of y-intercept and slope are b = −30, 319,966,899,308.016 and m = 15,057,249,898.695312. A regression line has been plotted as shown in Fig. 3.25, and investment for the year 2021 is $140.85 billion.

vii. **Regression Line for Fracked Oil & Gas Investment Prediction:**

The Python code for the regression line for Fracked Oil and Gas investment is provided below:

```
from statistics import mean
import numpy as np

fracked_list = df["Fracked Oil & Gas"].tolist()

year_list = df["Year"].tolist()

ys = np.array(fracked_list, dtype=np.float64)
xs = np.array(year_list, dtype=np.float64)

def best_fit_slope_and_intercept(xs,ys):
```

```
m = (((mean(xs)*mean(ys)) - mean(xs*ys)) /
((mean(xs)*mean(xs)) - mean(xs*xs)))

b = mean(ys) - m*mean(xs)

return m,b

m, b = best_fit_slope_and_intercept(xs,ys)

print(m,b)
7205808243.515625 -14457566694269.389

regression_line = [(m*x)+b for x in xs]

import matplotlib.pyplot as plt
from matplotlib import style
style.use('ggplot')

plt.scatter(xs,ys,color='black', label='data')
plt.plot(xs, regression_line, color='brown', label='regression
line')
plt.legend(loc="lower right")
plt.title('Best-Fit Line for Fracked Oil & Gas')
plt.xlabel('Year')
plt.ylabel('US Billion Dollar')
plt.show()
```

The below code is to predict the Fracked Oil and Gas investment by various banks for the year 2023:

```
predict_x1 = 2023
predict_y1 = (m*predict_x1) + b
print(predict_y1)
119783382362.7207
```

After computing, the values of y-intercept and slope are $b = -14,457,566,694,269.389$ and $m = 7,205,808,243.515625$. A regression line has been plotted as shown in Fig. 3.26, and investment for the year 2023 is $119.78 billion.

3.3.2 Multivariate Regression and Prediction Result

Multivariate Regression is used to predict an output using multiple independent variables. In this study, seven different consequences are predicted resulting from the investment by the banks in the fossil fuel sectors discussed earlier. The process starts by importing Pandas, NumPy libraries, and the "linear model" function from the scikit-learn library to load the dataset. The dataset is trained using data points of independent variables, and subsequently, the regression coefficient and regression intercept are calculated to predict the final value of output.

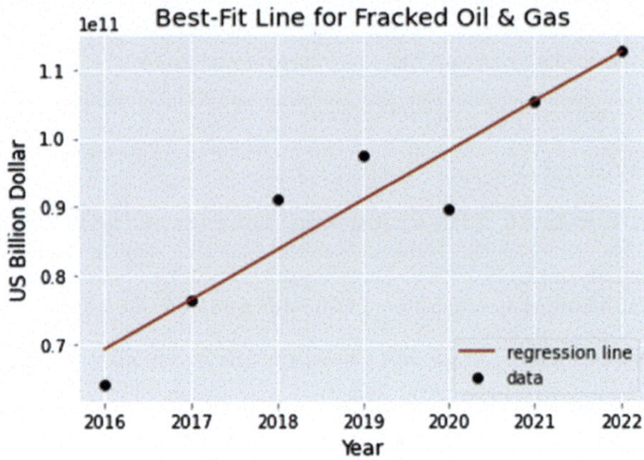

Fig. 3.26 Regression line for total fracked oil and gas investment by bank from the year 2016 to 2022

A. Methane Emission Prediction from Oil and Gas Sector

The Environmental Protection Agency (EPA) studying methane emissions from oil and gas production has found 90% higher emissions from oil production and 50% higher emissions for natural gas production [18]. After an in-depth study, it has been analyzed that methane emissions arise from Offshore Oil and Gas, Arctic Oil and Gas, and Fracked Oil and Gas extraction affecting climate change. Methane emission from oil and gas for the year 2023 has been predicted from the three sectors mentioned above. The snippet of the Python code used for predicting methane emission is shared below:

```
import pandas as pd
import numpy as np
from sklearn import linear_model

df = pd.read_excel(r"C:\Users\Acer\Downloads\Climate Change\Banks &
Climate            Change\Dataset\climate_change_main_dataset.xlsx",
sheet_name='Sheet11')
df

reg = linear_model.LinearRegression()
reg.fit(df[['Arctic Oil and Gas', 'Offshore Oil and Gas', 'Fracked Oil
and Gas']],df.Methane_Emission_from_Oil_and_Gas)

reg.coef_
array([ 4.42689700e-09, -2.16363086e-10, -1.28993676e-10])
reg.intercept_
1673.5220170638388

reg.predict([[15234196933,1.4085E+11,1.19783E+11]])
array([1695.03624764])
```

The methane emission for the year 2023 has been estimated as 1695.04 MMTCO_2e (million metric tons of CO_2 Equivalent) which is higher than the previous years. Table 3.1 clearly shows that methane emissions from oil and gas may increase due

Table 3.1 Predicted methane emission from the oil and gas sector

Year	Methane emission from oil and gas ($MMTCO_2e$)	Arctic oil and gas ($)	Offshore oil and gas ($)	Fracked oil and gas ($)
2016	1668.5	3,902,166,063	61,999,730,744	64,119,602,653
2017	1672.33	3,485,538,977	36,202,044,790	76,468,192,130
2018	1676.16	4,946,171,057	31,935,295,542	91,023,769,971
2019	1679.98	8,452,426,835	84,585,815,177	97,554,802,074
2020	1683.81	10,413,048,734	1.13094E + 11	89,605,338,898
2021	1687.484155	11,636,466,293	1.10735E + 11	1.05372E + 11
2022	1691.260237	13,435,331,613	1.25792E + 11	1.12578E + 11
2023	**1695.036248**	**15,234,196,933**	**1.4085E + 11**	**1.19783E + 11**

to investment in Offshore Oil and Gas, Arctic Oil and Gas, and Fracked Oil and Gas sectors.

B. **Methane Emission Prediction from Coal Mining Sector**

Due to the high-intensity mining of coal and coal power, methane emissions are rising, particularly from coal-related sectors. Using coal mining and coal power investment values, future methane emission from coal has been predicted. The Python code used to predict methane emission is stated as follows:

```
import pandas as pd
import numpy as np
from sklearn import linear_model

df = pd.read_excel(r"C:\Users\Acer\Downloads\Climate Change\Banks &
Climate          Change\Dataset\climate_change_main_dataset.xlsx",
sheet_name='Sheet12')
df

reg = linear_model.LinearRegression()
reg.fit(df[['Coal Power','Coal Mining']],df.Methane_Emission_from_
Coal_Mines)

reg.coef_
array([ 1.36330875e-10, -5.20781808e-10])

reg.intercept_
965.9387493333413

reg.predict([[40157015299,29715251865]])
array([955.9382278])
```

The methane emission from the coal mining sector for the year 2023 is estimated as 955.94 MMTCO2e (million metric tons of CO2 Equivalent) which is lower than the previous years. From Table 3.2, it can be concluded that methane emissions due to investments in the coal mining sector may reduce slightly with a low margin.

Table 3.2 Predicted methane emission from coal mining sector

Year	Methane emission from coal mines	Coal power	Coal mining
2016	961.02	43,653,998,108	20,360,341,416
2017	960.15	38,447,085,566	22,954,895,314
2018	959.28	47,925,761,321	26,247,697,916
2019	958.4	43,095,345,416	24,596,708,506
2020	957.53	39,052,037,833	25,352,298,379
2021	957.2733367	41,068,147,439	27,390,106,442
2022	956.6057822	40,612,581,369	28,552,679,154
2023	**955.9382278**	**40,157,015,299**	**29,715,251,865**

C. Global Temperature Prediction from All the Fossil Fuel Sectors

According to National Oceanic and Atmospheric Administration (NOAA) Annual Climate Report 2020, the combined land and ocean temperature has increased at an average rate of 0.13 °F (0.08 °C) per decade since 1880. However, the average rate of increase since 1981, 0.32 °F (0.18 °C), has been more than twice that rate [19]. In order to predict the global temperature for the year 2023, all fossil fuel sectors, namely, Tar Sands, Coal Mining, Coal Power, LNG, Arctic Oil and Gas, Offshore Oil and Gas, and Fracked Oil and Gas, have been considered. The below Python code is used to predict the global temperature:

```
import pandas as pd
import numpy as np
from sklearn import linear_model

df = pd.read_excel(r"C:\Users\Acer\Downloads\Climate Change\Banks &
Climate Change\Dataset\climate_change_main_dataset.xlsx",
sheet_name='Sheet13')
df

reg = linear_model.LinearRegression()
reg.fit(df[['Tar Sands','Coal Power','Coal Mining', 'LNG',
'Arctic Oil and Gas', 'Offshore Oil and Gas', 'Fracked Oil and
Gas']],df.Global_Temperature)

reg.coef_
array([ 2.88131463e-04, 1.69181804e-04, 2.89561733e-03,
1.22447931e-03, -1.60164242e-03, -2.12585782e-07, -1.24067589e-04])

reg.intercept_
-92140435.42085823

reg.predict([[8031034110,40157015299,29715251865,29628530242,
15234196933,140849645752.601,119783382362.72]])
array([0.99440867])
```

Table 3.3 Predicted global temperature from the fossil fuel sectors

Year	Global temperature	Tar sands	Coal power	Coal mining	LNG	Arctic oil and gas	Offshore oil and gas	Fracked oil and gas
2016	1.01	2.12E + 10	4.37E + 10	2E + 10	2.8E + 10	3.902E + 09	6.2E + 10	6.41E + 10
2017	0.92	4.3E + 10	3.84E + 10	2.3E + 10	1.8E + 10	3.486E + 09	3.6E + 10	7.65E + 10
2018	0.85	1.46E + 10	4.79E + 10	2.6E + 10	1.9E + 10	4.946E + 09	3.2E + 10	9.1E + 10
2019	0.98	2.22E + 10	4.31E + 10	2.5E + 10	2.7E + 10	8.452E + 09	8.5E + 10	9.76E + 10
2020	1.02	1.62E + 10	3.91E + 10	2.5E + 10	2.9E + 10	1.041E + 10	1.1E + 11	8.96E + 10
2021	0.97999942	1.42E + 10	4.11E + 10	2.7E + 10	2.7E + 10	1.164E + 10	1.1E + 11	1.05E + 11
2022	0.98798438	1.11E + 10	4.06E + 10	2.9E + 10	2.9E + 10	1.344E + 10	1.3E + 11	1.13E + 11
2023	**0.99440867**	**8.03E + 09**	**4.02E + 10**	**3E + 10**	**3E + 10**	**1.523E + 10**	**1.4E + 11**	**1.2E + 11**

The global temperature for the year 2023 is estimated as 0.994 °C, which is higher than the previous years. It is clear from Table 3.3 that global temperature may rise due to investment in the fossil fuel sectors.

D. **Mean Surface Temperature Prediction from All the Fossil Fuel Sectors**

In earth science, Global Surface Temperature (GST), sometimes referred to as Global Mean Surface Temperature (GMST), is calculated by averaging the temperature at the surface of the sea and air temperature over land [20]. Fossil fuel sectors like Tar Sands, Coal Mining, Coal Power, LNG, Arctic Oil and Gas, Offshore Oil and Gas, and Fracked Oil and Gas have been considered to predict the mean surface temperature. The Python code used to predict the mean surface temperature is as follows:

```
import pandas as pd
import numpy as np
from sklearn import linear_model
df = pd.read_excel(r"C:\Users\Acer\Downloads\Climate Change\Dataset\climate_change_main_dataset.xlsx", sheet_name='Sheet4')

df

reg = linear_model.LinearRegression()
reg.fit(df[['Tar Sands','Coal Power','Coal Mining','LNG','Arctic Oil and Gas','Offshore Oil and Gas','Fracked Oil and Gas']],df.Mean_Surface_Temperature)
```

```
reg.coef_
array([     1.22188096e-03,      7.17433409e-04,      1.22795042e-02,
5.19268897e-03, -6.79212151e-03, -9.05032006e-07, -5.26132979e-04])

reg.intercept_
-390741274.97792804

reg.predict([[8031034110,40157015299,29715251865,29628530242,
15234196933,140849645752.601,119783382362.72]])
array([1.15725183])
```

The mean surface temperature for the year 2023 is estimated as 1.15 °C, which is higher than in previous years. Table 3.4 shows that the mean surface temperature is rising due to the investment in all the fossil fuel sectors mentioned above.

E. **CO_2 Emission Prediction from All the Fossil Fuel Sectors**

Carbon dioxide (CO_2) makes up the vast majority of GHG emissions from the sector. CO_2 is released during the combustion of fossil fuels, such as coal, oil, and natural gas [21]. All fossil fuel sectors, namely, Tar Sands, Coal Mining, Coal Power, LNG, Arctic Oil and Gas, Offshore Oil and Gas, and Fracked Oil and Gas, discussed in this chapter, impact CO_2 emissions. Hence, all the seven fossil fuel sectors have been included while predicting CO_2 emissions for the year 2023. The Python code used to predict the CO_2 emission is as follows:

```
import pandas as pd
import numpy as np
from sklearn import linear_model

df = pd.read_excel(r"C:\Users\Acer\Downloads\Climate Change\Banks &
Climate          Change\Dataset\climate_change_main_dataset.xlsx",
sheet_name='Sheet15')
df

reg = linear_model.LinearRegression()
reg.fit(df[['Tar   Sands','Coal   Power','Coal   Mining',   'LNG',
'Arctic Oil and Gas', 'Offshore Oil and Gas', 'Fracked Oil and
Gas']],df.CO2_Avg_PPM)

reg.coef_
array([     1.66875976e-06,      9.80803396e-07,      1.67687189e-05,
7.08966849e-06, -9.27479054e-06, -9.11464726e-10, -7.18564231e-07])

reg.intercept_
-533215.3164395131

reg.predict([[8031034110,40,157,015,299,29,715,251,865,29,628,530,
242,15,234,196,933,140,849,645,752.601,119,783,382,362.72]])
array([421.43040745])
```

The CO_2 emission for the year 2023 has been estimated as 421.43 PPM. It is evident from Table 3.5 that CO_2 emissions may rise due to investments in the fossil fuel sectors.

Table 3.4 Predicted mean surface temperature for the fossil fuel sectors

Year	Mean surface temperature (°C)	Tar sands ($)	Coal power ($)	Coal mining ($)	LNG ($)	Arctic oil and gas ($)	Offshore oil and gas ($)	Fracked oil and gas ($)
2016	0.94	2.12E + 10	4.37E + 10	2.036E + 10	2.8E + 10	3.9E + 09	6.2E + 10	6.41E + 10
2017	0.84	4.3E + 10	3.84E + 10	2.295E + 10	1.8E + 10	3.5E + 09	3.62E + 10	7.65E + 10
2018	0.8	1.46E + 10	4.79E + 10	2.625E + 10	1.9E + 10	4.9E + 09	3.19E + 10	9.1E + 10
2019	0.88	2.22E + 10	4.31E + 10	2.46E + 10	2.7E + 10	8.5E + 09	8.46E + 10	9.76E + 10
2020	1.16	1.62E + 10	3.91E + 10	2.535E + 10	2.9E + 10	1E + 10	1.13E + 11	8.96E + 10
2021	1.06799754	1.42E + 10	4.11E + 10	2.739E + 10	2.7E + 10	1.2E + 10	1.11E + 11	1.05E + 11
2022	1.11593376	1.11E + 10	4.06E + 10	2.855E + 10	2.9E + 10	1.3E + 10	1.26E + 11	1.13E + 11
2023	**1.15725183**	**8.03E + 09**	**4.02E + 10**	**2.972E + 10**	**3E + 10**	**1.5E + 10**	**1.41E + 11**	**1.2E + 11**

Table 3.5 Predicted CO_2 emission for the fossil fuel sectors

Year	CO_2 average PPM	Tar sands	Coal power	Coal mining	LNG	Arctic oil and gas	Offshore oil and gas	Fracked oil and gas
2016	404.4125	2.12E + 10	4.365E + 10	2.036E + 10	2.8E + 10	3.9E + 09	6.2E + 10	6.41E + 10
2017	406.7575	4.3E + 10	3.845E + 10	2.295E + 10	1.8E + 10	3.5E + 09	3.6202E + 10	7.65E + 10
2018	408.715	1.46E + 10	4.793E + 10	2.625E + 10	1.9E + 10	4.9E + 09	3.1935E + 10	9.1E + 10
2019	411.6566667	2.22E + 10	4.31E + 10	2.46E + 10	2.7E + 10	8.5E + 09	8.4586E + 10	9.76E + 10
2020	414.2375	1.62E + 10	3.905E + 10	2.535E + 10	2.9E + 10	1E + 10	1.1309E + 11	8.96E + 10
2021	416.5205833	1.42E + 10	4.107E + 10	2.739E + 10	2.7E + 10	1.2E + 10	1.1074E + 11	1.05E + 11
2022	418.9754999	1.11E + 10	4.061E + 10	2.855E + 10	2.9E + 10	1.3E + 10	1.2579E + 11	1.13E + 11
2023	421.4304075	8.03E + 09	4.016E + 10	2.972E + 10	3E + 10	1.5E + 10	1.4085E + 11	1.2E + 11

F. **Minimum Arctic Sea Ice Extent Prediction**

The Arctic Sea ice extent is minimum on the day in a given year when Arctic Sea ice reaches its smallest extent, usually by the end of the summer melting season in September. Sea ice extent is an alternative measurement and is usually defined as the area with at least 15% of sea ice cover [22]. Arctic Sea ice melts due to the increasing temperatures and CO_2 emissions that get accumulated at the poles. Therefore, while predicting the minimum arctic sea ice extent, factors that directly affect it have been considered, which are "change in mean surface temperature", "global temperature", and "CO_2 emissions". The snippet of the code in Python for predicting minimum Arctic Sea ice extent is given below:

```
import pandas as pd
import numpy as np
from sklearn import linear_model

df = pd.read_excel(r"C:\Users\Acer\Downloads\Climate Change\Banks &
Climate           Change\Dataset\climate_change_main_dataset.xlsx",
sheet_name='Sheet16')
df

reg = linear_model.LinearRegression()
reg.fit(df[['Global  Temperature','Mean  Surface  Temperature','CO2
Average PPM']],df.Minimum_Arctic_Sea_Ice_Extent)

reg.coef_
array([-2.01169198, -0.94408673, -0.04311068])
```

Table 3.6 Predicted minimum arctic sea ice extent

Year	Minimum arctic sea ice extent	Global temperature	Mean surface temperature	CO_2 average PPM
2016	4.53	1.01	0.94	404.4125
2017	4.82	0.92	0.84	406.7575
2018	4.79	0.85	0.8	408.715
2019	4.36	0.98	0.88	411.657
2020	3.92	1.02	1.16	414.2375
2021	3.98	0.98	1.07	416.52
2022	3.82	0.99	1.12	418.98
2023	**3.66**	**0.99**	**1.16**	**421.43**

```
reg.intercept_
24.91849828927179
reg.predict([[0.99440867,1.15725183,421.4304075]])
array([3.65735844])
```

The estimated minimum Arctic Sea ice extent for the year 2023 is 3.66 million square kilometers. From Table 3.6, it can be seen that the minimum Arctic Sea ice extent has been decreasing continuously and is predicted to come down drastically. This is a major cause of concern as the melting of Arctic ice will result in an increase in sea level.

G. Mean Sea Level Prediction

Global warming is causing the global mean sea level to rise in two ways. First, glaciers and ice sheets worldwide are melting and adding water to the ocean. Second, the volume of the ocean is expanding as the water warms due to the increase in global temperature and surface temperature [23]. Therefore, factors that directly affect the mean sea level, like "Arctic Sea ice extent", "global temperature", and "surface temperature", have been considered for predicting the mean sea level. The snippet of the code in Python for predicting mean sea level is provided below:

```
import pandas as pd
import numpy as np
from sklearn import linear_model

df          =          pd.read_excel(r"C:\Users\Acer\Downloads\Climate
Change\Dataset\climate_change_main_dataset.xlsx",
sheet_name='Sheet6')
df

reg = linear_model.LinearRegression()
reg.fit(df[['Minimum Arctic Sea Ice Extent']],df.Mean Sea Level)

reg.coef_
array([-12.95757397])

reg.intercept_
```

Table 3.7 Predicted increase in mean sea level

Year	Increase in mean sea level	Minimum arctic sea ice extent	Global temperature	Mean surface temperature	CO_2 average PPM
2016	54.88	4.53	1.01	0.94	404.4125
2017	52.78	4.82	0.92	0.84	406.7575
2018	55.26	4.79	0.85	0.8	408.715
2019	61.93	4.36	0.98	0.88	411.657
2020	64.74	3.92	1.02	1.16	414.2375
2021	66.53431071	3.98227504	0.98	1.07	416.52
2022	69.30632909	3.8151227	0.99	1.12	418.98
2023	**72.16088532**	**3.65735844**	**0.99**	**1.16**	**421.43**

```
116.01976167638165

reg.predict([[3.98227441]])
array([64.41914644])
```

The estimated mean sea level for the year 2023 is 72.16 mm. From Table 3.7, it can be seen that the mean sea level may increase due to higher "CO_2 emissions", "global temperature", "surface temperature", and lower "minimum Arctic Sea ice extent".

3.3.3 *Performance Evaluation*

After regression analysis and predicting the results, it is time to evaluate the model. The model has been assessed using MSE and R^2 values. The Python code for evaluating the MSE and R^2 values is given below:

```
import pandas as pd
import numpy as np
from sklearn import linear_model

from sklearn.metrics import mean_squared_error
mean_squared_error(y_test,y_test_predicted)

from sklearn.metrics import r2_score
r2_score(y_test,y_test_predicted)
```

MSE should be in the range of 0–infinity, and the value closer to zero suggests that the model is a good predictor. R^2 value should lie between the range 0 and 1, and the value closer to 1 indicates a better fit for the model. Table 3.8 clearly indicates that the desired results in terms of MSE and R^2 have been achieved.

Table 3.8 Mean squared error and R^2 values

Prediction model	MSE	R^2
Methane emission from oil and gas	0.28	1.00
Methane emission from coal mines	0.26	0.89
Global temperature	0.00	1.00
Mean surface temperature	0.00	1.00
CO_2 Avg PPM	0.00	1.00
Minimum arctic sea ice extent	0.00	0.99
Mean sea level	0.001	0.99

3.4 Conclusions

The multivariate regression analysis has been carried out to predict the consequences related to climate change arising due to the investment by banks in fossil fuel sectors from 2016 to 2020. The analysis shows that the climate change consequences (dependent variables) and the investments by banks in different fossil fuel sectors (independent variables) are correlated. It has been analyzed that climate change consequences like methane emissions, carbon emissions, global temperature, mean surface temperature, mean sea level, and minimum arctic sea ice extent are adversely impacted due to the investment by banks in Tar Sands, Coal Power, Coal Mining, LNG, Fracked Oil and Gas, Offshore Oil and Gas, and Arctic Oil and Gas.

From this research, it can be concluded that banks should phase out their investment in fossil fuel sectors in a strategic manner and play a vital role towards achieving the net-zero target. There must be more disclosure of financial information by the banks, investors, and companies targeting carbon-intensive sectors. The consequences of climate change, viz. sea level, global temperature, sea ice extent, and GHG emissions, are at an all-time high level, and the banks, regulators, and governments must act now.

The dataset of investment by the top 60 worldwide banks during the year 2016–2020 has been used in this research. The scope of this study can further be extended with the larger dataset, and the results of the prediction can be used in formulating worldwide investment policies in fossil fuel sectors or carbon-intensive sectors/projects. Disclosures in the financial sector are expected to increase, and there is a promising path for future research.

References

1. Coal Importance to the World (2021) Society for Mining, Metallurgy & Exploration (SME). https://www.smenet.org/What-We-Do/Technical-Briefings/Coal-s-Importance-in-the-US-and-Global-Energy-Supp
2. Bian Z, Inyang HI, Daniels JL, Otto F, Struthers S Environmental issues from coal mining and their solutions, Mining Science and Technology (China). https://doi.org/10.1016/S1674-5264(09)60187-3

3. https://www.ucsusa.org/resources/what-are-tar-sands
4. Wykes S, Heywood S (2010) Tar sands fuelling the climate crisis, undermining EU energy security and damaging development objectives. Urbaniak D, de Clerck P (eds). https://www.banktrack.org/download/tar_sands_fuelling_the_climate_crisis_undermining_eu_energy_security_and_damaging_development_objectives
5. Gautier DL, Bird KJ, Charpentier RR, Grantz A, Houseknecht DW, Klett TR, Moore TE, Pittman JK et al (2009) Assessment of undiscovered oil and gas in the arctic. Science 324:1175–1179
6. Harsem Ø, Eide A, Heen K (2011) Factors influencing future oil and gas prospects in the arctic. Energy Policy 39(12):8037–8045. ISSN 0301-4215, https://doi.org/10.1016/j.enpol.2011.09.058
7. http://www.exploration-production-services.de/en/o-offshore.html
8. Offshore Oil and Gas Production. https://www.planete-energies.com/en/medias/close/offshore-oil-and-gas-production#:~:text=Offshore%20oil%20and%20gas%20production%2C%20which%20involves%20extracting%20oil%20and,to%20the%20related%20environmental%20impacts.&text=Offshore,-production%20accounts%20for
9. Burkett V (2011) Global climate change implications for coastal and offshore oil and gas development. Energy Policy 39(12):7719–7725. ISSN 0301-4215, https://doi.org/10.1016/j.enpol.2011.09.016. https://www.sciencedirect.com/science/article/pii/S0301421511007051
10. https://www.nrdc.org/stories/fracking-101
11. https://www.shell.com/energy-and-innovation/natural-gas/liquefied-natural-gas-lng.html
12. Pavlenko N, Comer B, Zhou Y, Clark N, Rutherford D (2020) The climate implications of using LNG as a marine fuel, Working paper 2020–02. International Council on Clean Transportation
13. Sailing to nowhere: Liquified Natural Gas is not an effective climate strategy. https://www.nrdc.org/sites/default/files/sailing-nowhere-liquefied-natural-gas-report.pdf
14. Banks and Liquified Natural Gas (LNG) https://www.banktrack.org/campaign/banks_and_liquefied_natural_gas_lng
15. https://www.wellbeingatschool.org.nz/information-sheet/understanding-and-interpreting-box-plots
16. https://chartio.com/learn/charts/what-is-a-scatter-plot/
17. https://corporatefinanceinstitute.com/resources/excel/study/correlation-matrix/
18. https://www.seas.harvard.edu/news/2021/03/oil-and-natural-gas-production-emit-more-methane-previously-thought
19. https://www.climate.gov/news-features/understanding-climate/climate-change-global-temperature
20. https://en.wikipedia.org/wiki/Global_surface_temperature
21. https://www.epa.gov/ghgemissions/sources-greenhouse-gas-emissions#:~:text=Carbon%20dioxide%20(CO2)%20makes,natural%20gas%2C%20to%20produce%20electricity
22. https://en.wikipedia.org/wiki/Arctic_sea_ice_decline
23. https://www.climate.gov/news-features/understanding-climate/climate-change-global-sea-level#:~:text=Global%20warming%20is%20causing%20global,expanding%20as%20the%20water%20warms

Chapter 4
Effect of Non-renewable Energy Sources on Climate Change in India—Literature Review and Data Preparation

4.1 Introduction

The United Nations has mentioned that climate change is the "defining crisis of our time". After more than 10,000 years of relative stability, which is the full span of human civilization, the earth's climate is changing at an alarming rate [1]. The climate of the earth has always been fluctuating due to its natural drivers but what is happening in today's time is something which is catastrophic. Recent variations and changes have caused untimely disasters like storms, droughts, forest fires, heat waves, and rising sea levels which are changing the present face of the earth at high speeds. This is and will continue to pose a grave threat to the existence of life on earth if appropriate action is not taken immediately.

The dependence of humankind on the energy sector has been long known. As the Industrial Revolution started, dependence on energy grew exponentially, and countries started exploiting resources, mostly fossil fuels, to boost economic growth and development. The consequences of such practices were not considered at that time, but those implications have now started to show their effects. The burning of fossil fuels results in the release of gases such as carbon dioxide, sulfur dioxide, and methane which are also known as Greenhouse Gases (GHGs). The problem with these gases is that they trap the heat reflected by the earth's surface and prevent it from escaping the atmosphere. The inability of heat to escape from earth's atmosphere causes the temperatures to rise and can have numerous irreversible effects to such an extent which puts humankind as well as the environment in great danger. Data suggests that fuel combustion for the generation of electricity has been the primary cause of the release of GHGs in the atmosphere [2]. Figure 4.1 shows the contribution of various sectors to GHG emissions (in $MtCO_2e$—metric tons of carbon dioxide equivalent) in India from 1999 to 2018. The plot has been prepared on the basis of the dataset fetched from the website of "Climate Watch Data" [3]. Over the years, fossil fuel burning to generate electricity has grown rapidly to meet the exponential growth in consumption demand. Data shows that the global Total Primary Energy Supply (TPES) doubled between 1971 and 2008 [2]. The problem is aggravated as most of

N. Sharma and P. K. De, *Towards Net-Zero Targets*, Advances in Sustainability Science and Technology, https://doi.org/10.1007/978-981-19-5244-9_4

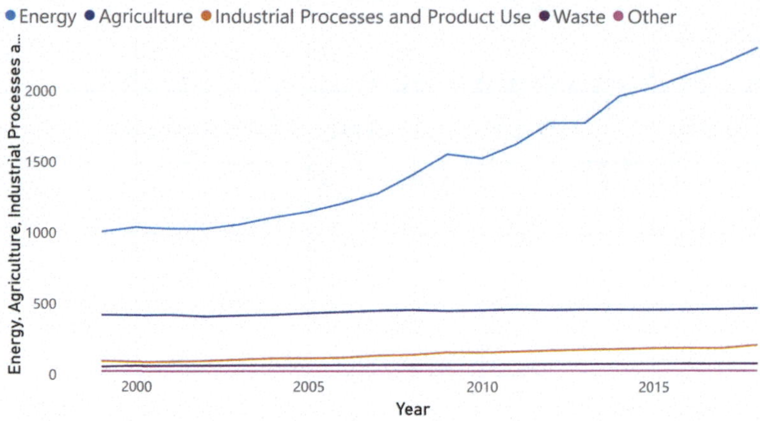

Fig. 4.1 Sector-wise greenhouse gas emissions (in $MtCO_2e$) for India

the nations are too focused on developing their economies through industrialization, which requires energy. But in the process, they pay little attention to the consequences of this growth in energy consumption. The industrial boom for countries such as China has been linked to exponential growth in energy consumption [2]. The situation in India is equally alarming and is clearly evident in Fig. 4.2, which has been prepared on the basis of the dataset extracted from the International Energy Agency (IEA) [4]. The exponential growth shown in Fig. 4.2 represents the fact that the energy industry and climate change have a direct and strong relationship. Because of this strong interdependence, the options to tackle this are very limited. Fortunately, fossil fuels are not the only source of energy. In order to mitigate the impact of the energy sector on climate change, there is a strict and urgent need to shift to cleaner renewable sources of energy such as hydropower, solar power, geothermal energy, bioenergy, etc.

The evolution of energy is depicted in Fig. 4.3. Energy, when defined in terms of physics, is the capacity or ability to do the work [5]. It can exist in various forms like kinetic, potential, thermal, chemical, electrical, nuclear, magnetic, and many more such forms. According to the law of conservation of energy, it is stated that energy cannot be created or destroyed but can be transformed into different forms [5]. Today's civilization has come to its existence because human beings have learned to convert energy and make it useful for various purposes. The energy is stored in different forms and can be generated from many sources. The two main categories of available energy sources are renewable and non-renewable energy sources. Renewable sources of energy can be easily replenished, whereas non-renewable sources of energy cannot be refilled. These are the primary sources that are used to generate energy that humankind consumes for its survival and also used to generate secondary sources like electricity.

The energy industry or energy sector of the world is involved in all the activities which are related to the production, conversion, distribution, supply, refining,

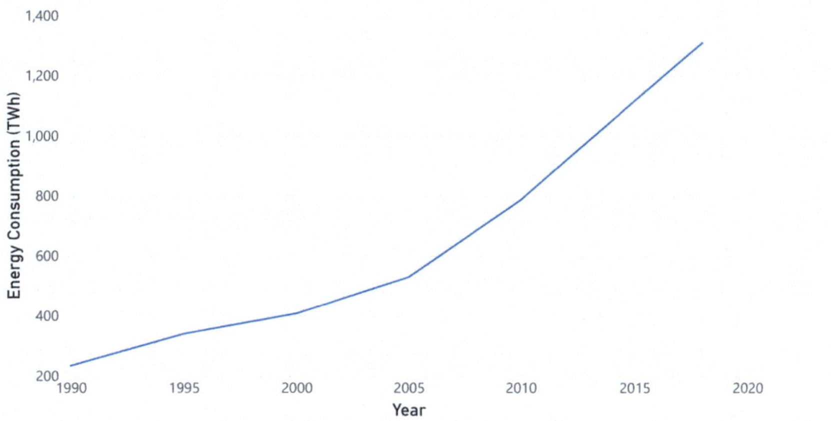

Fig. 4.2 Energy consumption (in TWh—terawatt-hour) for India

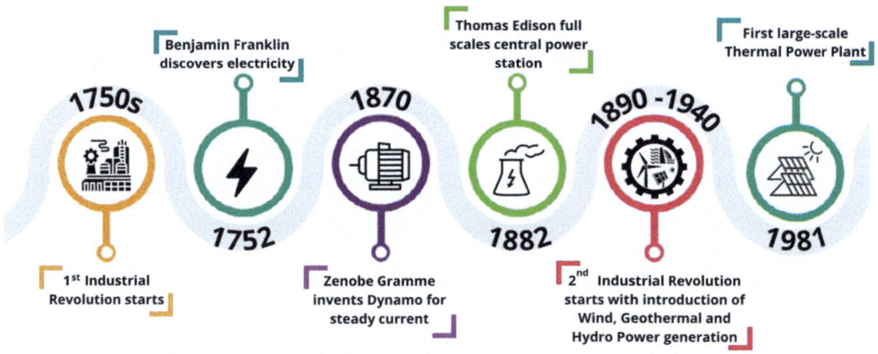

Fig. 4.3 Evolution of the energy sector

and consumption of energy. The sector includes companies which are related to the production of energy and the process of production, for example, through non-renewable sources like fossil fuels, coal, oil, natural gas etc. or renewable sources like hydroelectric power, thermal power, wind power, and many other sources.

The dependence of humankind on energy and its use has been key in the development of society. Although electricity was discovered in 1752 by Benjamin Franklin, it was not until the late nineteenth century that the world started to grasp the implications, potential, and consequences of this form of energy [6]. The history of power generation is very much unclear, being marked by contributions by hundreds of people. Perhaps the most important moment came in 1870, when a Belgian inventor, Zenobe Gramme, came up with a dynamo that produced a steady direct current suitable for powering motors—a discovery that generated a sense of enthusiasm and excitement about electricity's potential to light and power the world [6]. Not much later, in 1882, Thomas Edison achieved his vision of a full-scale central power station with a system of conductors to distribute electricity [6].

Major strides in alternating current technology opened up a new horizon for power generation. Hydropower marked several milestones between 1890 and 1900 [6]. Energy generation really took off during the second Industrial Revolution, which gave rise to a number of industries dependent on electricity. Installation of power generation plants became a common phenomenon. Electrification allowed the final major developments in manufacturing methods during the Second Industrial Revolution.

The growing dependence and need for energy led to larger plants being set up and more fossil fuels being burnt. This growing dependence during the twentieth century on carbon-emitting sources of energy such as fossil fuels and carbon-emitting renewables such as biomass brought the energy industry under scrutiny for being a major contributor to pollution, and consequently climate change. Until recently, fossil fuels were the main source of energy generation in most parts of the world.

4.2 Literature Review

In this section, an attempt has been made to provide an overview of the existing literature related to the topic of study. For easy understanding, the section has been divided into two parts—Systematic Literature Review and Contemporary Literature Review.

4.2.1 Systematic Literature Review

To learn more about the work and research which have been done in the domains of climate change, energy and forecasting methods, a systematic literature review is performed to explore these areas and to understand the importance as well as the association of climate change and the energy sector. Three digital libraries, namely, Springer, IEEE Xplore, and Elsevier, have been explored to get the data specific to research work done in the above domains. The search criteria used are "Climate Change", "Climate Change and Energy", and "Climate Change, Energy and Forecasting". Figures 4.4, 4.5, and 4.6 depict that the number of publications on climate change is highest with a continuous increasing trend in all the digital libraries. Elsevier has the highest number of publications compared to Springer and IEEE Xplore in all the search criteria. It is clearly visible from Figs. 4.4, 4.5, and 4.6 that the work done in the domain of "Climate change, Energy and Forecasting" is significantly low compared to other search criteria. On reviewing all the three digital libraries together, it is seen that there has been a jump of 226% in search criteria "Climate Change" in the last decade (2011–2020). Likewise, it is observed that there is an increase of 248% in the search criteria "Climate Change and Energy" and an increase of 278% in the search criteria "Climate Change, Energy and Forecasting".

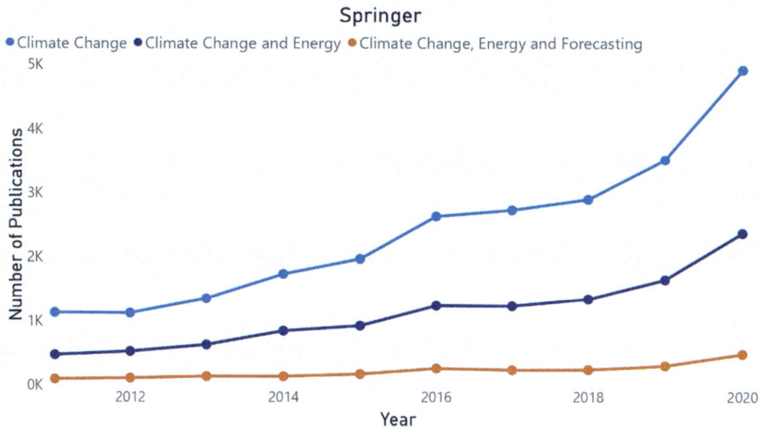

Fig. 4.4 Systematic literature review of Springer

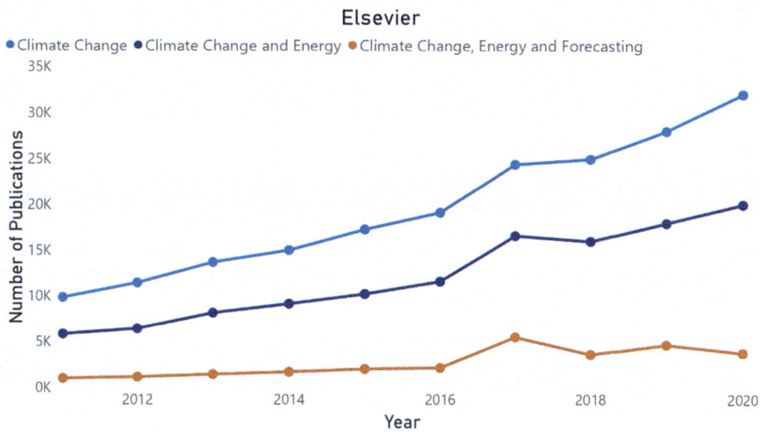

Fig. 4.5 Systematic literature review of Elsevier

There has been a significant rise in the publication in the area of Climate Change and the Energy Sector. Hence, it can be inferred that this is a matter of utmost importance and has been gaining rapid interest among the researchers.

4.2.2 Contemporary Literature Review

This sub-section presents a comparison of various literature in terms of objectives, issues addressed, aspects covered, techniques, and procedures applied in the

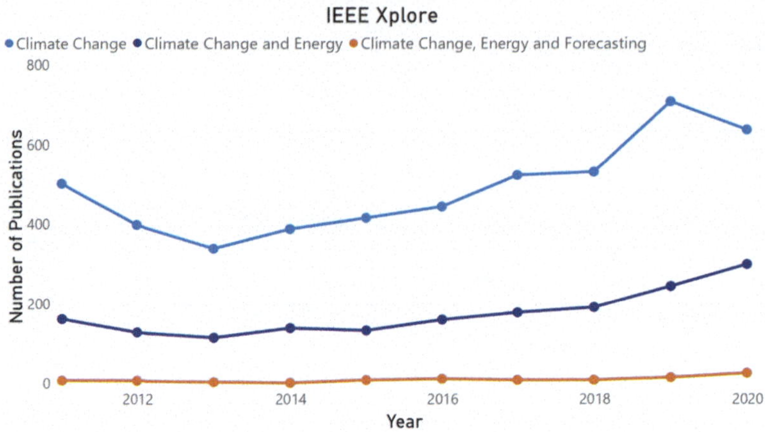

Fig. 4.6 Systematic literature review of IEEE Xplore

domain of the energy sector and its impact on climate change. The summary of the contemporary literature review is presented in Table 4.1.

Climate change is a threatening environmental concern of today's time. The energy sector, being a major contributor to climate change, has come under a lot of monitoring. Energy companies have been facing tough challenges in reducing GHG emissions through the replacement of current technologies and investments in sustainable development. Climate change mitigation is a global measure taken through regulations and standardization. The need to meet growing demands for energy, 80% of which is obtained from fossil fuel, is the main cause of climate change, as confirmed by the 4th IPCC report. Climate mitigation strategies focus on implementing steps that allow development by respecting the need to reduce emissions through various agreements that have been established at the global level. Rio de Janeiro convention in 1992 aimed at setting emission reduction targets, where the EU decided to reduce emissions by 20% [7]. Europe's 2020 strategy aims to focus on smart, sustainable, and inclusive development with the following targets: (a) reduce emissions by 20% by using clean technologies, (b) increase the proportion of renewable energy to 20% of total consumption, and (c) increase energy efficiency by 20% [7]. Steps towards achieving these targets will be a major stride towards reducing the impact of climate change.

The Fifth Assessment Report from the Intergovernmental Panel on Climate Change (IPCC) briefs about the rigorous scientific basis of the original source material. According to the report, climate change is indisputable, and human activities, particularly carbon dioxide emissions, are very likely to be the dominant cause [8]. Models used in the IPCC reports show that there will be a wide range of changes in the GHG emissions in the twenty-first century, and with this rate, the global average temperature is projected to rise 2.6–4.8 degrees Celsius higher than the present [8]. It is stated that even if the emissions are stopped completely by today's date, the temperature will continue to rise due to the presence of past GHG emissions, which

Table 4.1 Contemporary literature review

S. No	Article/paper name	Author(s)	Year	Issues addressed	Aspects covered	TTechnologies—procedures applied/models built
1	Energy prices and turning points: the relationship between income and energy use/carbon emissions	Amy K, Richmond, Robert K. Kaufinann	2006	Economic growth and reduction in energy use or emissions that cause global climate change	Relationship between GDP and energy use	14 Models were used to test whether the environmental Kuznets curve (EKC) can describe the relationship between GDP and energy used
2	The contribution of energy consumption to climate change: a feasible policy direction	Usenobong F. Akpan, Godwin E. Akpan	2012	Increased global use of fossil fuels and consequently its effect on world energy supply and consumption	Electricity, heat, and transport sector—mitigation policies and mechanisms	Comparative study of emissions from different sources of energy in various world regions and countries
3	Climate change: implications for the energy sector	A report by the University of Cambridge and the World Energy Council	2014	Challenges faced by the energy sector due to climate change	– Increase in emissions from the energy sector due to an increasing trend in consumption – Achieving a reduction in the GHG emissions – Implications faced by the energy sector due to the political action at a global level – Challenge in investment in the low-carbon technologies	– Key findings from the Intergovernmental Panel on Climate Change Fifth Assessment Report – Earth System Model (ESM) Climate Model – Integrated Assessment Models – Atmosphere–Ocean General Circulation Models (AOGCMs)

(continued)

Table 4.1 (continued)

S. No	Article/paper name	Author(s)	Year	Issues addressed	Aspects covered	TTechnologies—procedures applied/models built
4	The energy sector and the climate change challenge	Laman Mammadzada, Umid Abuzarli	2015	– The challenge for energy sector is represented by climate change – Outlined relevant issues from an economic point of view	Fitting oil-related energy consumption within economic development	Case Study—Report—Presented some issues from the oil and gas industry
5	A review of renewable energy sources, sustainability issues and climate change mitigation	Owusu P. A., Sarkodie S. A.	2016	– Opportunities associated with renewable energy sources – Energy security, energy access, social and economic development – Climate change mitigation and reduction of environmental and health impacts	– Various forms of renewable energy such as bioenergy, hydropower, geothermal energy, solar energy, wind energy, and tidal energy – Scope of sustainability in these various forms and discuss the shortcomings – Statistics to clearly help understand the depth of the problem	A theoretical research-based study on different renewable sources of energy

(continued)

Table 4.1 (continued)

S. No	Article/paper name	Author(s)	Year	Issues addressed	Aspects covered	TTechnologies—procedures applied/models built
6	Accelerating SDG7 achievement policy briefs in support of the First SDG7 review at the UN high-level political forum 2018 Policy brief #15 interlinkages between energy and climate change	United Nations Department of Economic and Social Affairs	2018	– Progress on achieving climate goals – To reduce the emission gap while also meeting energy access for all and air-quality goals	– Energy sector as a major factor contributing the climate change – Emission gap – Progress in promoting sustainable energy sources, renewable energy sources – Promoting technology and mitigation plans	Reports developed by United Nations Framework Convention on Climate Change (UNFCCC), International Renewable Energy Agency (IRENA), The European Commission and the UN Environment
7	Tackling climate change with machine learning	Priya L. Donti, Anna Waldman-Brown	2019	– Climate mitigation and adaptation – Meta-level tools that enable climate-related strategies	Electricity systems, industry production and energy	Theoretical study
8	Envisioning energy futures in the North Atlantic oil industry: avoidance, persistence, and transformation as responses to climate change	Mark C.J. Stoddarta, Patrick McCurdyb, Natalie Slawinskic, Cory G. Collinsa	2020	Energy futures and climate change in the oil sector	Events relating to issues of climate change and de-carbonization within the oil industry	Ethnography approach of research

(continued)

Table 4.1 (continued)

S. No	Article/paper name	Author(s)	Year	Issues addressed	Aspects covered	TTechnologies—procedures applied/models built
9	Changing climate risk in the UK: a multi-sectoral analysis using policy-relevant indicators	N. W. Arnell, A. L. Kay, A. Freeman, A. C. Rudd, J. A. Lowe	2021	– Risks across different sectors in the UK due to climate change projections – Resilience and policies required to achieve net-zero emissions	Sectors included in the study are: agriculture, transport, energy, health, infrastructure, water	Policy guidelines based on UKCP18 climate projections

are already present in the atmosphere [8]. The key findings of the report are that the trend of the energy demand at a global level is growing, leading to an increase in the emissions from the energy sector, which is posing challenges for the sector. Various measures can be followed to achieve a reduction in GHG emissions. At the same time, several challenges can be faced due to the political action at the global level and investment in low-carbon technologies [8]. Therefore, it is clear that the energy sector is impacting the climate in a big way and also getting affected due to the change. These challenges, in turn, are posing threats to thermal power plants, nuclear plants, solar and wind energy generation, the coal industry, power lines, the oil and gas sector, and the hydropower generation sector. It also affects the production and transmission of various energy sources [8]. The report highlights resilience options and mitigation solutions that can be adopted by the energy sector. Some of these are fuel extraction and conversion, renewables, fuel switching, increasing efficiency of transmission, carbon capture and bioenergy, reducing final energy demand, and policies towards strong energy and climate change mitigation [8].

The report on "Accelerating SDG 7 Achievement (policy brief 15 inter-linkages between energy and climate change)" shows the status of energy and climate along with the progress towards fulfilling the defined objectives and the primary actions required to be taken by the governments to achieve these [9]. The energy sector itself is responsible for two-thirds of GHG emissions worldwide, whereas, on the other hand, around 1.1 billion people around the world have no access to electricity or have access to unreliable sources of electricity networks. The report suggests that a proper balance of policies can help to achieve the goal of energy access around the world without trading off the climate change objectives and goals [9]. Progress and development in the use of renewable sources of energy have resulted in a positive outcome of stabilizing carbon emissions to a large extent since 2014. However, the rate of development and advancement of these sources are not enough to fight the climate change crisis which the world is facing at an alarming rate, and a new set of political agreements is needed to achieve the goal. Climate change poses challenges for the energy sector, thereby affecting it on a large scale. The report also discusses the current status of climate change and emphasizes the emission gap, which needs to be addressed for sufficient mitigation [9]. Several measures, such as promoting renewable energy sources, technologies, advancing energy efficiency, strengthening carbon pricing, limiting coal-based power, and ways through which governments can enhance the resilience of energy systems, have also been stated.

Akpan et al. believe that energy has been the main driving force for socio-economic development [2]. According to them, the overexploitation of energy by some nations to meet their growing needs has resulted in the negative impact of the source on the environment. One of the major concerns highlighted in this study is the global rise in temperature, which is the result of rapid urbanization and exponential growth in the human population, subsequently leading to a rapid increase in energy demand. To address this problem, the authors have focused on data from Australia, New Zealand, the Russian Federation, Japan, the United States, and most of the European countries. The study mainly attempts to address climate change from a historical perspective, the role of energy in climate change, and the options for mitigation [2]. The authors have suggested the measures that can be adopted, which include regularization and standardization of pollution output limits, the introduction of taxes and penalties, subsidies and incentives, research and development, etc. In addition, the authors suggest the government to invest in non-renewable energy and research and development.

Owusu et al. focus on the shift to renewable energy and the associated opportunities that come with this shift both at the global and the local level [10]. These include energy security, energy access, social and economic development, climate change mitigation, and reduction of environmental and health impacts. The study discusses various forms of renewable energy such as bioenergy, hydropower, geothermal energy, solar energy, wind energy, and tidal energy. It shows the scope of sustainability in these various forms as well as discusses some of their shortcomings like market failures, lack of information, access to raw materials for future renewable resource deployment, and the daily carbon footprint. It forms a link between renewable energy and its impact on climate change. Some statistics have been presented to help clearly understand the depth of the problem at hand. The benefits of different types of renewable energy sources have been discussed in detail. The shift to cleaner sources of energy not only affects the environmental factors but greatly influences the social as well as the economic factors of the area. These impacts of various forms of energy have been discussed at length. But as discussed above, all these investments in cleaner energy sources can only be helpful if they can be sustained in the long run. Hence, it can be said that renewable energy has a very clear and direct relationship with sustainability because neither can exist without the other. Different opportunities for sustainable development have been explained with the help of renewable energy [10]. To address these concerns, some policy measures and recommendations arising from the study can help to an extent. Investment and research in these fields can be of great importance as they help to come up with better as well as cheaper technologies. These measures, along with international cooperation at a large scale, aimed at reducing carbon footprint, and can greatly help to solve the challenges that arise with non-renewable sources of energy.

Donti et al. emphasize on different domains where Machine Learning (ML) can play a very important role and bear fruitful results in combating climate change which has now become a crisis for humanity. In recent years, ML has become a very broadly used and powerful tool and can be used to address this climate risk [11]. Electricity production and consumption are solely responsible for about 25% of GHG emissions produced due to human activities. In order to reduce GHG emissions from the electricity sector, there is a need to move to low-carbon sources, reduce emissions from existing plants, and implement the changes across all the countries [11]. A few suggested measures that have been discussed in the study are supply and demand forecasting, optimizing the electricity system goals using deep neural networks, developing new materials to harness energy from natural sources, and improving the scheduling and flexible demand by using optimization solutions like neural networks, genetic algorithms, etc. ML can be used to process the existing data and physics to figure out the structure of new material. Supervised learning, active learning, and generative models can be used to synthesize, model, and design material. Graphical models can be used to detect faults, forecast electricity prices, and help locate the ideal location for the installation of wind turbines, dams, etc. Besides, advanced ML like experimental design, decision trees, and deep learning techniques can be used to identify and manage sites for geothermal energy and to accelerate the process of nuclear energy production [11].

Climate change is a grave threat that the world is facing today. Around 97 percent of the scientists studying climate have stated that human activity is the main driver in this crisis. With the boom of AI technology, it has come out to be a very strong weapon in combating climate change, such as smart grid design, developing low-emission infrastructure, and modeling climate change predictions. It has now become an urgent requirement to consider AI in the climate crisis [12].

4.2.3 Non-renewable Sources of Energy in India

India draws its energy from various non-renewable and renewable sources of energy. The major amount of power and energy supplied all across the country is produced by non-renewable sources, which primarily include coal, oil, and natural gas. This sub-section presents the statistics pertaining to non-renewable sources.

A. **Coal Reserves in India**

Coal is the most abundant and widely used source of energy in India. India holds the 5th rank and accounts for 9% of the total world's coal reserves [13]. With the mining, extraction, and consumption of coal for the past 220 years, starting from 1774, India is now left with around 111 years of coal as of 2016, provided the present rate of consumption is kept constant [13, 14]. Figure 4.7 and Table 4.2 show the coal reserves of India till the year 2020 in million tons [15].

The biggest coal reserve available in India is in the state of Jharkhand, followed by Odisha and Chhattisgarh.

B. **Crude Oil Reserves in India**

India ranks 24th position in the world for holding oil reserves and accounts for around 0.3% of the world's total reserves [13]. The oil and gas industry is among the top 8 sectors in India and plays an important role in robust economic growth [16]. India, with its current rate of consumption, has approximately 3 years of oil left as of 2016 [13]. Figure 4.8 and Table 4.3 show the oil reserves of India till the year 2020 in million tons [15].

The top crude oil reserve is available in the state of Assam, followed by Gujarat.

C. **Natural Gas Reserves in India**

As of 2017, India has 22 years of natural gas left in its reserves which account for 1% of the total world's natural gas reserves making India hold the 22nd rank position

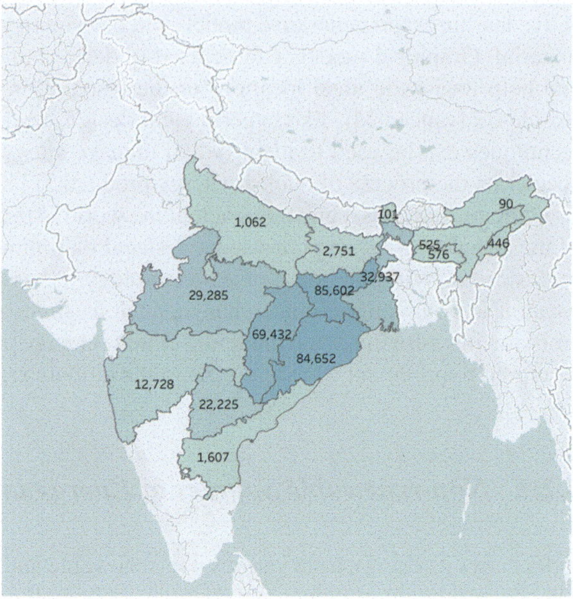

Fig. 4.7 Coal reserves of India (as of 2020)

Table 4.2 Coal reserves of India (as of 2020)

States/ UTs	Total	
	01.04.2019	01.04.2020
Andhra Pradesh	1,607	1,607
Arunachal Pradesh	90	90
Assam	525	525
Bihar	1,834	2,751
Chhattisgarh	59,908	69,432
Jharkhand	84,506	85,602
Madhya Pradesh	28,793	29,285
Maharashtra	12,677	12,728
Meghalaya	576	576
Nagaland	446	446
Odisha	80,840	84,652
Sikkim	101	101
Uttar Pradesh	1,062	1,062
West Bengal	31,690	32,937
Telangana	21,839	22,225
All India Total	**326,495**	**344,021**

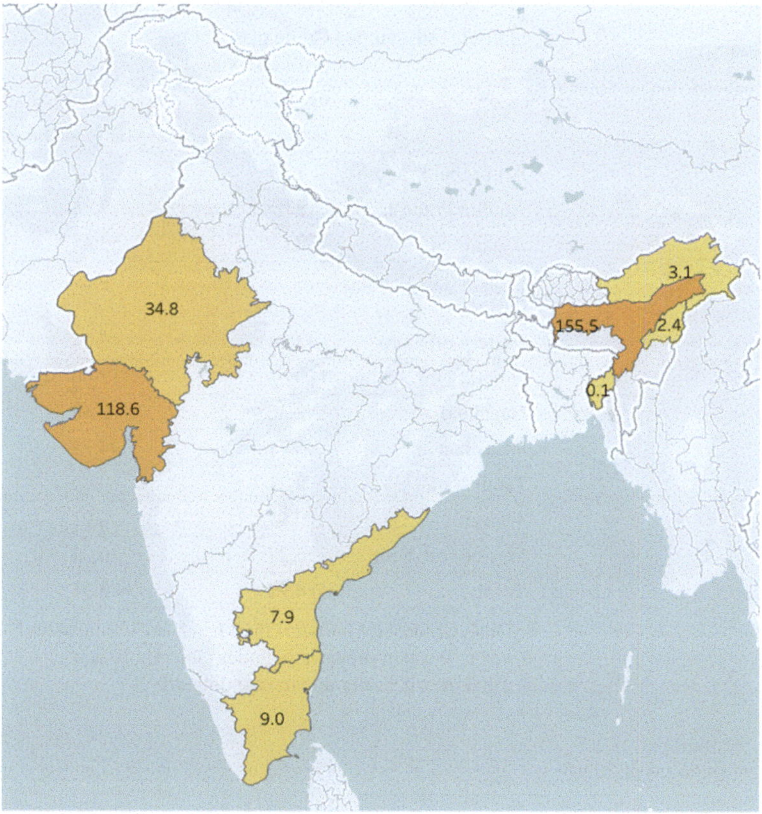

Fig. 4.8 Oil reserves of India (as of 2020)

[13]. Figure 4.9 and Table 4.4 show the natural gas reserves of India dated till the year 2020 in billion cubic meters [15].

The leading gas reserve is available in Eastern Offshore, followed by Western Offshore.

4.2.4 Electricity Generation, Land Temperature, Emissions in India

This sub-section presents the statistics pertaining to electricity generation, GHG emissions, GHG emissions from the energy sector, fugitive emissions, and land temperature.

Table 4.3 Oil reserves of India (as of 2020)

States/UTs/Region	Crude petroleum (million tonnes)	
	01.04.2019	01.04.2020
	Estimated reserves	Estimated reserves
Arunachal Pradesh	2.67	3.12
Andhra Pradesh	8.05	7.85
Assam	158.62	155.45
Cold bed methane (CBM)[a]	–	–
Eastern offshore	42.34	42.05
Gujarat	119.63	118.60
Nagaland	2.38	2.38
Rajasthan	40.71	34.78
Tamil Nadu	9.21	9.03
Tripura	0.07	0.07
Western offshore	235.27	230.04
Total	**618.95**	**603.37**

[a] CBM: Cold Bed Methane (Jharkhand, West Bengal and M.P.)
Notes 1. Western offshore includes Gujarat offshore
2. Total may not tally due to rounding off

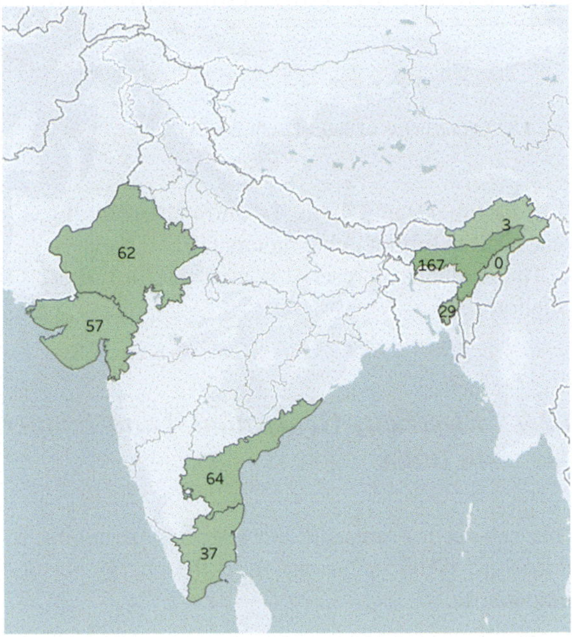

Fig. 4.9 Natural gas reserves of India (as of 2020)

Table 4.4 Natural gas reserves of India (as of 2020)

States/UTs/Region	Natural gas (billion cubic metres)	
	01.04.2019	01.04.2020
	Estimated reserves	Estimated reserves
Arunachal Pradesh	1.64	2.72
Andhra Pradesh	63.57	63.85
Assam	170.71	166.60
Cold bed methane (CBM)[a]	72.56	72.23
Eastern offshore	565.15	556.33
Gujarat	59.45	57.13
Nagaland	0.09	0.09
Rajasthan	63.03	61.80
Tamil Nadu	38.00	37.09
Tripura	23.24	29.45
Western offshore	323.19	324.60
Total	**1380.63**	**1371.89**

[a] CBM: Cold Bed Methane (Jharkhand, West Bengal and M.P.)
Notes 1. Western offshore includes Gujarat offshore
2. Total may not tally due to rounding off

A. Generation of Electricity in India

Figure 4.10 shows the trend of "Total Electricity Generation" and generation of electricity by different sources like "Coal", "Gas", "Hydroelectricity", "Solar Energy", and "Wind Energy" in India from 1990 to 2016. The production is calculated in terawatt-hour (TWh), and the data is collected from the website "Our World in Data" [17]. The following figure is developed using Power BI.

Figure 4.10 shows that the trend of electricity generation has been increasing over 27 years, and the coal sector is contributing to the maximum generation.

B. Total GHG Emissions in India

Figure 4.11 shows the trend of "Total GHG Emissions", "Emissions by Energy Sector", and "Emissions by Electricity and Heat" in India from 1990 to 2016. The emissions are calculated in million tons, and the data is collected from the website of "Our World in Data" [18]. The following figure is generated using Power BI.

It is evident from Fig. 4.8 that the total GHG emissions have been increasing over the period of 27 years.

C. GHG Emissions from Energy Sector in India

Figure 4.12 shows the trend of different GHG emissions from the energy sector in India during the period 1990–2016. The emissions are calculated in metric tons,

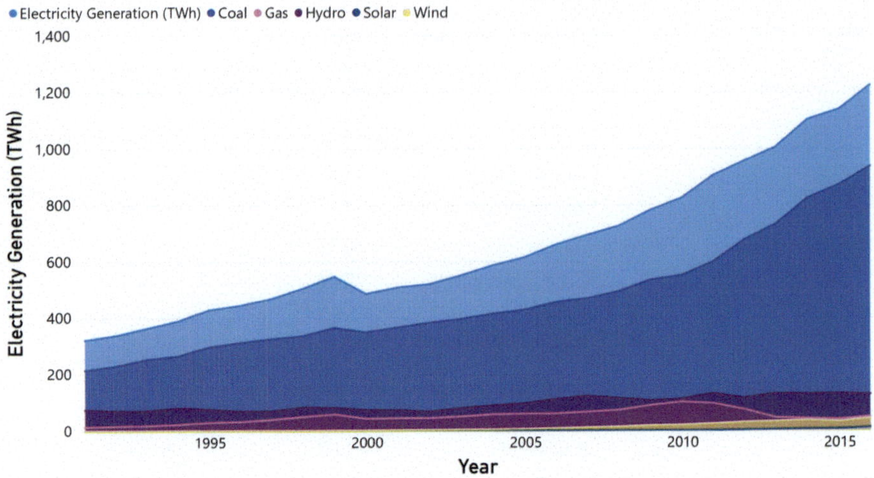

Fig. 4.10 Total electricity generation and electricity generation by various sources

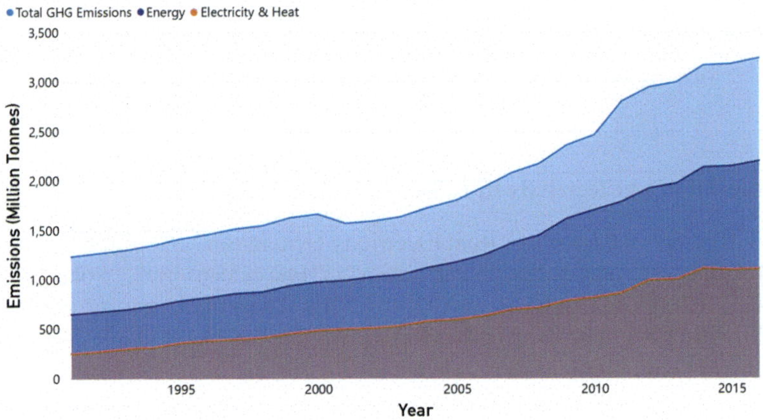

Fig. 4.11 GHG emissions from energy sector and electricity and heat

and the data is collected from the website "Our World in Data" [19]. These majorly include carbon dioxide (CO_2), methane (CH_4), and nitrogen dioxide (NO_2). It can be observed from Fig. 4.9 that the trends of methane and nitrogen dioxide emissions have been consistent over the specified span of time. In contrast, the trend of carbon dioxide emissions has a significant rise over the years.

D. **Total Fugitive Emissions in India**

Figure 4.13 shows the trend of different fugitive GHG emissions from the energy sector in India from 1990 to 2016. These are the types of emissions which are due to leakages and irregular releases. The emissions are calculated in metric tons, and the

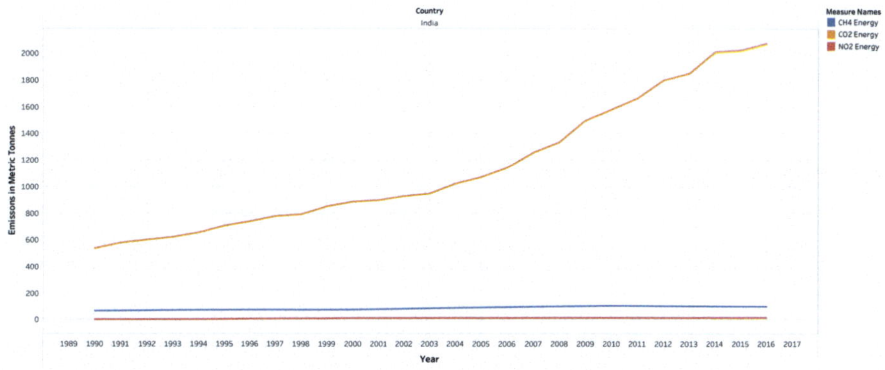

Fig. 4.12 GHG emissions from the energy sector of India

data is collected from the website "Our World in Data" [20]. These majorly include carbon dioxide (CO_2), methane (CH_4), and nitrogen dioxide (NO_2). From the figure, it can be observed that fugitive emissions of nitrogen dioxide have been constant, whereas the fugitive emissions of carbon dioxide have decreased during the period 1990–2003 and then have remained constant over time. It can further be observed that the fugitive emissions of methane gas have witnessed significant growth over the years till 2010, and from then, it has been decreasing at a slower rate.

E. **Average Land Temperature in India**

Figure 4.14 shows the "Average Land Temperature" trend in India from 1990 to 2016. The measure is recorded in Celsius, and the data is collected from the website "data.world" [21]. Power BI is used to create the following graph, which shows an overall increasing trend line.

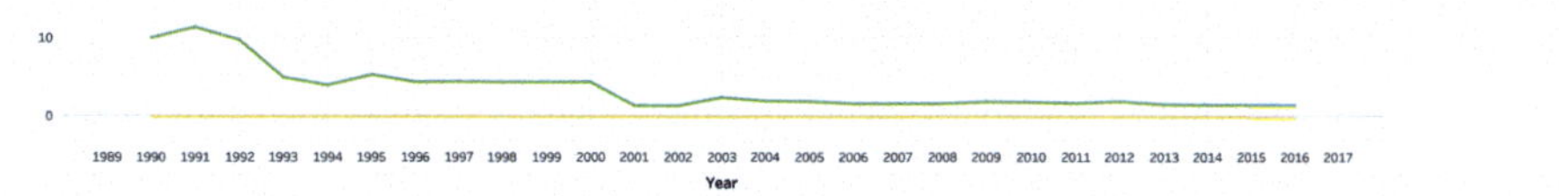

Fig. 4.13 Fugitive GHG emissions

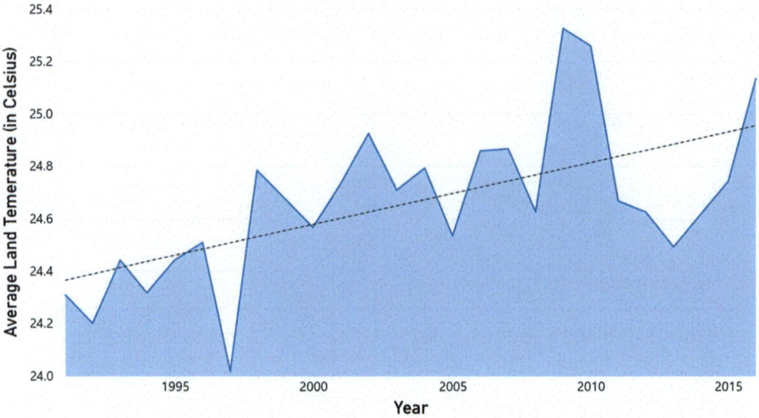

Fig. 4.14 Average land temperature

4.2.5 Problem Statement

India is a developing country and one of the fastest growing economies in the world. It is the home to the second largest population. With a tremendous rate of growth in industrialization and urbanization, India's economy is reaching new heights every day [22]. India's power sector is considered to be the most diversified in the world [22]. It generates power and energy from a spectrum of different sources like coal, oil, and natural gas to renewable sources such as hydro, wind, solar, and nuclear energy. Energy generation, demand, distribution, and consumption are accelerating at high speed and are expected to increase rapidly in the coming years. The energy industry is continuously redefining itself to meet the demands and production of various sectors of the economy.

Along with the exponential growth which the energy sector is going through, it has a major task to maintain the climate change policies issued by the government. Figure 4.1 shows that the energy sector has been the largest contributor to GHG emissions in India. Moreover, there is negligible work that has been carried out to handle the climate imbalance created due to the energy sector by using machine learning, which is also evident from the literature survey. It is a difficult and very challenging goal that India needs to work on by balancing advancement, progress, and development as well as the climate change crisis.

The aim of the research is to study India's energy sector and different sources of non-renewable energy and to clearly distinguish their potential and contributions towards climate change. Besides, an attempt is made to predict trends in temperature due to emissions from all the resources.

4.3 Material and Methods

This section talks about data source identification focusing on selected data sources and converting them into one final dataset and its description. In addition, this section also presents the algorithms used to build predictive models.

4.3.1 Data Source Identification

A huge amount of datasets capture GHG emissions, which can be categorized in terms of sector, fuel, and region. Along with this, plenty of datasets can be found on average land temperatures for all the regions of the world since 1700. After a thorough examination of all the datasets, two datasets have been chosen for the analysis in this research.

A. **Data Source**

A dataset from the "Our World in Data" website has been identified to carry out the research related to GHG emissions from different sources [21]. Likewise, the dataset from Kaggle [23], which in turn has been prepared using the dataset from Berkeley Earth [24], is used to study the increasing temperatures and the average land temperatures caused by climate change over a specific period of time.

B. **Dataset Description and Attributes**

The dataset that has been identified for the research represents CO_2 emissions (in tons) produced from different fuel sources for each country from the year 1750–2016. The fuel sources include coal, gas, cement, flaring, oil, and other industries. Since the present study focuses on the impact generated by the energy sector, only energy-related sources, i.e., oil, flaring, coal, and gas, have been considered. After removing irrelevant variables, data for India from 1962 to 2016 has been extracted, as the observations before this period for all four sources are not available. Table 4.5 shows the complete dataset of carbon emissions from different sources.

The dataset for "Average Land Temperatures" represents the monthly average temperature in degree Celsius for several countries from 1750 to 2013. Out of these observations, temperatures for India have been extracted. The dataset has the month-wise "temperature" data which is aggregated for 12 months to compute the average temperature for any particular year. The next step is to join this dataset with the "CO_2 emissions" dataset. The "CO_2 emissions" dataset has the data until 2016, whereas the "Average Land Temperatures" dataset has the data up to the year 2013 only. Therefore, the data for the remaining 3 years has been fetched from the Data Website of India [25]. Now both the datasets have been combined to get one single dataset containing CO_2 emissions from different fuel sources along with average yearly temperatures from 1962 to 2016, as shown in Table 4.6.

Table 4.7 presents the description of the attributes of the final dataset.

Table 4.5 CO_2 emissions (in tons) from different fuel sources

Year	Oil	Flaring	Coal	Gas
1962	24,885,252	43,967	103,455,048	10,992
1963	27,106,272	135,568	110,628,323	10,992
1964	28,121,885	476,332	106,123,937	21,985
1965	30,814,921	751,137	116,675,620	282,134
1966	35,676,368	831,728	117,204,446	311,440
1967	32,697,536	1,454,608	119,320,098	487,312
1968	42,249,584	1,370,336	124,059,556	751,120
1969	45,350,200	1,260,440	123,600,275	996,627
1970	49,332,096	1,421,632	123,783,890	923,328
1971	55,442,660	1,417,943	127,205,858	1,084,525
1972	58,877,807	1,212,804	134,630,117	1,205,476
1973	61,749,392	1,447,280	138,068,690	1,168,816
1974	60,628,208	1,725,744	145,913,540	1,359,344
1975	61,838,227	2,011,565	161,642,124	1,733,097
1976	63,438,496	1,817,344	169,238,849	2,293,664
1977	68,557,104	2,092,144	177,844,271	2,480,528
1978	75,317,184	1,982,224	175,017,866	2,740,672
1979	82,663,504	2,216,720	180,962,947	3,030,128
1980	83,176,464	1,165,152	197,708,812	2,487,856
1981	89,713,040	3,176,688	210,701,636	2,883,568
1982	94,406,624	3,814,224	213,524,719	4,360,160
1983	99,169,824	4,316,192	232,957,495	5,243,184
1984	106,104,774	5,199,167	231,929,030	6,170,118
1985	116,759,685	5,953,949	253,766,418	7,199,698
1986	121,857,312	4,964,720	273,686,281	10,310,496
1987	127,671,122	6,628,126	292,406,108	12,776,272
1988	133,615,088	7,470,896	317,769,553	15,205,600
1989	150,714,021	11,002,922	340,265,710	18,517,739
1990	158,166,612	9,870,757	367,204,537	21,097,187
1991	166,422,544	7,833,632	393,783,556	23,563,184
1992	188,054,800	3,565,072	412,190,058	27,256,496
1993	186,878,656	3,700,640	433,937,762	26,996,352
1994	198,954,243	4,056,028	456,576,692	29,000,420
1995	216,281,275	3,290,257	472,611,857	39,050,735
1996	243,160,345	3,715,280	505,449,329	39,746,906
1997	244,201,936	3,568,736	528,716,964	46,023,504

(continued)

Table 4.5 (continued)

Year	Oil	Flaring	Coal	Gas
1998	262,892,000	3,312,256	525,113,381	47,595,360
1999	300,174,406	3,004,480	561,584,294	42,315,536
2000	311,884,068	3,110,736	573,956,040	43,872,736
2001	315,679,793	3,224,343	583,180,029	43,568,933
2002	324,133,911	2,744,336	599,725,807	47,133,696
2003	329,448,168	2,029,856	627,069,857	49,987,952
2004	347,001,949	1,901,610	671,230,692	49,584,753
2005	351,121,200	1,685,440	722,828,836	50,438,624
2006	368,313,059	1,839,328	775,022,660	49,442,016
2007	390,403,142	2,418,240	844,065,308	52,695,648
2008	411,656,206	2,114,133	912,512,326	63,046,600
2009	430,854,802	1,905,276	985,801,868	112,684,767
2010	437,563,102	1,864,976	1,018,777,120	134,244,240
2011	458,049,457	2,077,488	1,078,259,360	136,364,239
2012	487,647,302	2,077,488	1,225,745,910	125,604,798
2013	494,707,949	2,121,456	1,318,073,822	110,689,227
2014	510,232,333	1,674,448	1,447,232,990	109,467,500
2015	553,834,416	1,674,448	1,473,541,370	106,870,633
2016	612,136,920	1,883,293	1,541,435,738	113,622,467

4.3.2 Algorithms

This study employs two forecasting models in order to predict future temperature trends based on the previous values.

A. **ARIMA Model**

ARIMA stands for autoregressive integrated moving average and is one of the most widely used models in time series forecasting. In an autoregressive model such as the ARIMA model, the forecasting of the variable of interest is done by using a linear combination of the past values of the variable itself. The term autoregressive indicates that it is a regression of the variable against itself [26]. If we include differencing with autoregression and a moving average model, we get the ARIMA model, which can be represented as

$$y_t = \phi_1 y_{t-1} + \ldots \phi_p y_{t-p} - \theta_1 z_{t-1} - \ldots - \theta_q z_{t-q} + z_t$$

where z_t is a white noise process. This is known as an ARIMA (p, d, q) model, where

p: order of the autoregressive part,

Table 4.6 The combined dataset - CO2 emissions (in tons) and temperatures (degree Celsius)

Year	Oil	Flaring	Coal	Gas	Temp
1962	24,885,252	43,967	103,455,048	10,992	23.88475
1963	27,106,272	135,568	110,628,323	10,992	24.09575
1964	28,121,885	476,332	106,123,937	21,985	24.0705
1965	30,814,921	751,137	116,675,620	282,134	24.09091667
1966	35,676,368	831,728	117,204,446	311,440	24.43858333
1967	32,697,536	1,454,608	119,320,098	487,312	23.99208333
1968	42,249,584	1,370,336	124,059,556	751,120	23.86475
1969	45,350,200	1,260,440	123,600,275	996,627	24.50116667
1970	49,332,096	1,421,632	123,783,890	923,328	24.07408333
1971	55,442,660	1,417,943	127,205,858	1,084,525	23.634
1972	58,877,807	1,212,804	134,630,117	1,205,476	24.27216667
1973	61,749,392	1,447,280	138,068,690	1,168,816	24.38658333
1974	60,628,208	1,725,744	145,913,540	1,359,344	24.17575
1975	61,838,227	2,011,565	161,642,124	1,733,097	23.86216667
1976	63,438,496	1,817,344	169,238,849	2,293,664	24.19958333
1977	68,557,104	2,092,144	177,844,271	2,480,528	24.28891667
1978	75,317,184	1,982,224	175,017,866	2,740,672	24.00733333
1979	82,663,504	2,216,720	180,962,947	3,030,128	24.55358333
1980	83,176,464	1,165,152	197,708,812	2,487,856	24.59308333
1981	89,713,040	3,176,688	210,701,636	2,883,568	24.22841667
1982	94,406,624	3,814,224	213,524,719	4,360,160	24.21825
1983	99,169,824	4,316,192	232,957,495	5,243,184	23.938
1984	106,104,774	5,199,167	231,929,030	6,170,118	24.18508333
1985	116,759,685	5,953,949	253,766,418	7,199,698	24.48633333
1986	121,857,312	4,964,720	273,686,281	10,310,496	24.2945
1987	127,671,122	6,628,126	292,406,108	12,776,272	24.78908333
1988	133,615,088	7,470,896	317,769,553	15,205,600	24.72775
1989	150,714,021	11,002,922	340,265,710	18,517,739	24.16233333
1990	158,166,612	9,870,757	367,204,537	21,097,187	24.17816667
1991	166,422,544	7,833,632	393,783,556	23,563,184	24.31175
1992	188,054,800	3,565,072	412,190,058	27,256,496	24.22291667
1993	186,878,656	3,700,640	433,937,762	26,996,352	24.40866667
1994	198,954,243	4,056,028	456,576,692	29,000,420	24.29641667
1995	216,281,275	3,290,257	472,611,857	39,050,735	24.42975
1996	243,160,345	3,715,280	505,449,329	39,746,906	24.43691667
1997	244,201,936	3,568,736	528,716,964	46,023,504	24.04733333

(continued)

Table 4.6 (continued)

Year	Oil	Flaring	Coal	Gas	Temp
1998	262,892,000	3,312,256	525,113,381	47,595,360	24.75541667
1999	300,174,406	3,004,480	561,584,294	42,315,536	24.65983333
2000	311,884,068	3,110,736	573,956,040	43,872,736	24.498
2001	315,679,793	3,224,343	583,180,029	43,568,933	24.58191667
2002	324,133,911	2,744,336	599,725,807	47,133,696	24.913
2003	329,448,168	2,029,856	627,069,857	49,987,952	24.64991667
2004	347,001,949	1,901,610	671,230,692	49,584,753	24.63458333
2005	351,121,200	1,685,440	722,828,836	50,438,624	24.49541667
2006	368,313,059	1,839,328	775,022,660	49,442,016	24.73208333
2007	390,403,142	2,418,240	844,065,308	52,695,648	24.64991667
2008	411,656,206	2,114,133	912,512,326	63,046,600	24.406
2009	430,854,802	1,905,276	985,801,868	112,684,767	25.14666667
2010	437,563,102	1,864,976	1,018,777,120	134,244,240	25.05083333
2011	458,049,457	2,077,488	1,078,259,360	136,364,239	24.41558333
2012	487,647,302	2,077,488	1,225,745,910	125,604,798	24.64083333
2013	494,707,949	2,121,456	1,318,073,822	110,689,227	25.41325
2014	510,232,333	1,674,448	1,447,232,990	109,467,500	24.616667
2015	553,834,416	1,674,448	1,473,541,370	106,870,633	24.741667
2016	612,136,920	1,883,293	1,541,435,738	113,622,467	25.133333

Table 4.7 Attribute description of emissions and temperature dataset

Feature name	Description
Year	The year in which the observations were recorded
Oil	Average CO_2 emissions in tons from the source of oil
Flaring	Average CO_2 emissions in tons from the source of flaring
Coal	Average CO_2 emissions in tons from the source of coal
Gas	Average CO_2 emissions in tons from the source of gas
Temp	Average land temperature in India during the given year

d: degree of first differencing involved,

q: order of the moving average part.

B. **ARIMAX Model**

The ARIMAX model is simply an extension of the ARIMA model. It stands for autoregressive integrated moving average exogenous. In the ARIMA model, only univariate time series data is considered. But in the ARIMAX model, multivariate time series data is considered, and the variables which affect the target variables are known as the exogenous variables, hence the term exogenous [27]. An ARIMAX model simply adds the covariate on the right-hand side of the equation:

$$y_t = \beta x_t + \phi_1 y_{t-1} + \ldots \phi_p y_{t-p} - \theta_1 z_{t-1} - \ldots - \theta_q z_{t-q} + z_t$$

where x_t is a covariate at time t and β is its coefficient.

4.4 Conclusions

In this chapter, various research works in the domain of the energy sector and climate change have been studied from different perspectives. A review of diverse literature has been done for "different kinds of energy resources", "emissions from various non-renewable energy sources in India", and "interrelation between energy domain and climate change" to understand the importance of the net-zero goal. The literature review clearly highlights the risk of climate change, and intensive measures have to be taken towards a sustainable future. India derives its energy and power majorly from non-renewable sources, and with the increasing rate of usage, it will soon be depleted, which in turn will create a threat to millions of lives. Increasing emissions have resulted in higher average land temperatures due to extensive use of non-renewable energy sources for many decades. In the last section of this chapter, a dataset has been prepared to represent CO_2 emissions from various non-renewable sources of energy along with the average land temperatures of India for a period of 54 years. A detailed explanation of the forecasting models to be employed has also been discussed.

References

1. Climate risk and response: physical hazards and socioeconomic impacts (2020) McKinsey Global Institute. https://www.mckinsey.com/business-functions/sustainability/our-insights/climate-risk-and-response-physical-hazards-and-socioeconomic-impacts
2. Akpan FU, Akpan G (2015) The contribution of energy consumption to climate change: a feasible policy direction. https://www.researchgate.net/publication/227411028_The_Contribution_of_Energy_Consumption_to_Climate_ChangeA_Feasible_Policy_Direction
3. Historical Country Greenhouse Gas Emissions Data. Climate Watch. https://www.climatewatchdata.org/ghg-emissions
4. Energy Consumption, International Energy Agency. https://www.iea.org/data-and-statistics/data-browser/?country=WORLD&fuel=Energy%20consumption&indicator=CO2Industry
5. Energy, Britannica, last updated on 16 Nov 2021. https://www.britannica.com/science/energy

6. History of power: the evolution of the electric generation industry. Power Magazine 22 Dec 2020. https://www.powermag.com/history-of-power-the-evolution-of-the-electric-generation-industry/
7. Mammadzada L, Abuzarli U (2015) The Energy Sector and the Climate Change Challenge. In: International conference on competitiveness of agro-food and environmental economy proceedings, vol 4. The Bucharest University of Economic Studies, pp 340–345
8. Climate Change: Implications for the Energy Sector (2014) Intergovernmental panel on climate change fifth assessment report. University of Cambridge. https://www.cisl.cam.ac.uk/system/files/documents/IPCC_AR5__Implications_for_Energy__Briefing__WEB_EN.pdf
9. Accelerating SDG 7 Achievement Policy Brief 15: interlinkages between energy and climate change', United Nations Framework Convention on Climate Change (2018). https://sustainabledevelopment.un.org/content/documents/17498PB15.pdf
10. Owusu PA, Sarkodie SA (2016) A review of renewable energy sources, sustainability issues and climate change mitigation. https://www.tandfonline.com/doi/full/https://doi.org/10.1080/23311916.2016.1167990
11. Tackling Climate Change with Machine Learning. Cornell University, 10 June 2019. https://arxiv.org/abs/1906.05433
12. Dhar P (2020) The Carbon Impact of Artificial Intelligence. https://www.nature.com/articles/s42256-020-0219-9
13. India Energy, Worldometer. https://www.worldometers.info/energy/india-energy/
14. Ministry of Coal, Government of India, https://coal.gov.in/
15. Ministry of Statistics and Program Implementation. http://mospi.nic.in/publication/energy-statistics-india-2021
16. Oil and Gas Industry in India, India Brand Equity Foundation, last updated on 19 Nov 2021. https://www.ibef.org/industry/oil-gas-india.aspx
17. Ritchie H, Roser M (2020) Energy Production and Consumption. OurWorldInData. https://ourworldindata.org/energy-production-consumption
18. Ritchie H, Roser M (2020) CO_2 and Greenhouse Gas Emissions. OurWorldInData. https://ourworldindata.org/co2/country/india?country=~IND
19. Ritchie H, Roser M (2020) Greenhouse Gas Emissions. OurWorldInData. https://ourworldindata.org/greenhouse-gas-emissions
20. Ritchie H, Roser M (2020) CO_2 Emissions by Fuel. OurWorldInData. https://ourworldindata.org/emissions-by-fuel
21. Global Climate Change Data. Data World. https://data.world/data-society/global-climate-change-data
22. Power Sector in India, India Brand Equity Foundation, last updated on 19 Nov 2021. https://www.ibef.org/industry/power-sector-india.aspx
23. Climate Change: Earth Surface Temperature Data. Kaggle. https://www.kaggle.com/berkeleyearth/climate-change-earth-surface-temperature-data
24. Berkeley Earth. http://berkeleyearth.org/data/
25. All India Seasonal and Annual Mean Temperature Series (2016) Government of India, 23 Sept 2016. https://data.gov.in/catalog/all-india-seasonal-and-annual-mean-temperature-series
26. Autoregressive Models, Otexsts. https://otexts.com/fpp2/AR.html
27. Hyndman RJ (2020) The ARIMAX Model Muddle. https://robjhyndman.com/hyndsight/arimax/

Chapter 5
Using Machine Learning to Predict the Effect of Non-renewable Energy Sources on Climate Change in India

5.1 Introduction

The energy domain can be stated as one of the most important and crucial aspects of climate change which the world is facing today. In Chap. 4, the interrelation between energy and climate change has been explored and reviewed through different sources of literature. Some statistics have been presented to understand the relationship between the non-renewable sources of energy in India and CO_2 emissions. Besides, thorough research has been carried out to comprehend the energy resources of India, the average land temperatures, and its exponential growth during the past many years. The chapter also presented the sources used to fetch the data, methods adopted to prepare the final dataset, and a description of its various features. The rigorous study helped in framing the problem statement, which is to predict the temperature due to emissions from non-renewable energy sources.

This chapter provides the methods and procedures to implement the forecasting algorithms that have been discussed in the previous chapter. It has enabled the study to predict the future trend of India's "Average Land Temperatures" as a result of "CO_2 emissions" from non-renewable sources of energy. The chapter illustrates the step-by-step process of time series analysis that is used to forecast temperature values. The experimentation has been done using Python. The last section throws light on the energy sector and its impact on climate change. The chapter concludes by highlighting the measures that can be taken from an individual perspective as well collectively to fight the climate risk.

5.2 Exploratory Data Analysis (EDA)

Exploratory Data Analysis (EDA) is a method of analyzing data through the use of visual techniques. With the increased use of statistical summaries and graphical representations, EDA is used to find trends, patterns, and to test assumptions. In this

N. Sharma and P. K. De, *Towards Net-Zero Targets*, Advances in Sustainability Science and Technology, https://doi.org/10.1007/978-981-19-5244-9_5

section, numerous trends and relationships between variables have been described using exploratory data analysis. The exploratory data analysis is performed using the Plotly Python library [1]. It is an open-source interactive library. This is a very useful tool for data visualization that helps in analyzing the data quickly and easily. Plotly graph objects provide a high-level, user-friendly interface to Plotly. It can create scatter plots, line charts, bar charts, box plots, histograms, pie charts, and other graphs and charts. The variables described through visualizations in the analysis are "Gas", "Oil", "Flaring", "Coal" and "Temperature". The association of variables like gas, oil, flaring, and coal with the variable temperature can be observed from the scatter plot diagrams in Figs. 5.1, 5.2, 5.3 and 5.4. Snippets of the codes to create scatter plots are provided below:

```
import plotly.express as px
import pandas as pd

df = pd.read_excel(r"C:\Users\Aryam Sood\Fuel Wise OG.xlsx")

fig = px.scatter(df, x='Temp', y='Gas',labels=dict(Gas='Emissions
from Gas in Metric Tonnes',Temp='Temperature in degree Celsius'))
fig.show()
fig.show()

fig = px.scatter(df, x='Temp', y='Oil',labels=dict(Oil='Emissions
from Oil in Metric Tonnes',Temp='Temperature in degree Celsius'))
fig.show()
fig.show()

fig = px.scatter(df, x='Temp', y='Flaring',labels=dict(Flaring=
'Emissions from Flaring in Metric Tonnes',Temp='Temperature in degree
Celsius'))
fig.show()

fig = px.scatter(df, x='Temp', y='Coal',labels=dict(Coal='Emissions
from Coal in Metric Tonnes',Temp='Temperature in degree Celsius'))
fig.show()
fig.show()
```

All the figures indicate a significant relationship with "temperature". Gas, oil, and coal show a strong positive relationship which indicates an increase in temperature with the increase in emissions from these three variables. However, it is observed for the variable "Flaring" that the association is not strong with temperature compared to other variables.

The correlation between the variables has also been performed using Pearson correlation as shown in Fig. 5.5. The code used to generate the Pearson correlation matrix is mentioned as follows:

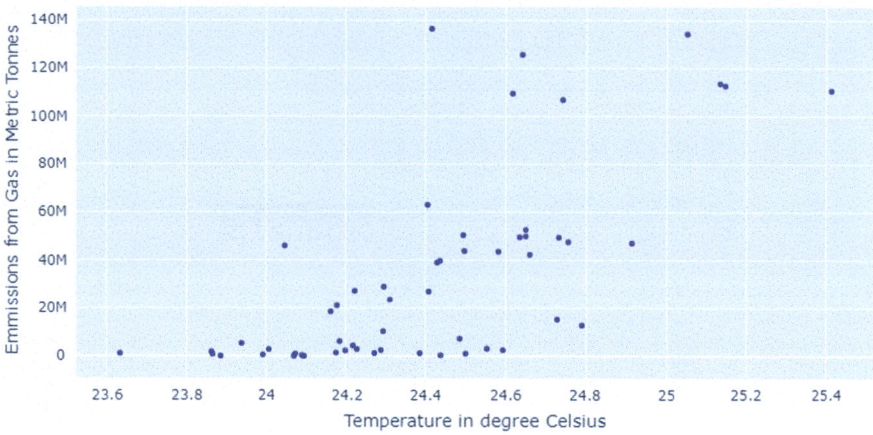

Fig. 5.1 Scatter plot of emissions from gas versus temperature

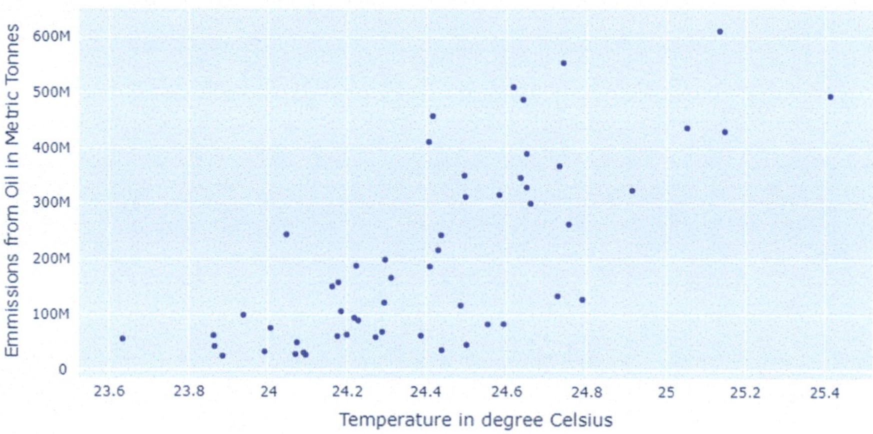

Fig. 5.2 Scatter plot of emissions from oil versus temperature

```
fig1 = px.imshow(df.corr(),color_continuous_scale='RdBu_r')
fig1.show()
```

The correlations between temperature and the variables, namely, "Oil", "Coal", and "Gas", are not only positive but also very strong, as observed in the scatter plot figures. The variables "Gas", "Oil", and "Coal" are highly correlated with each other. This does not hold good for the "Flaring" variable, which might be due to a drastic decrease in the emissions from Flaring over the years.

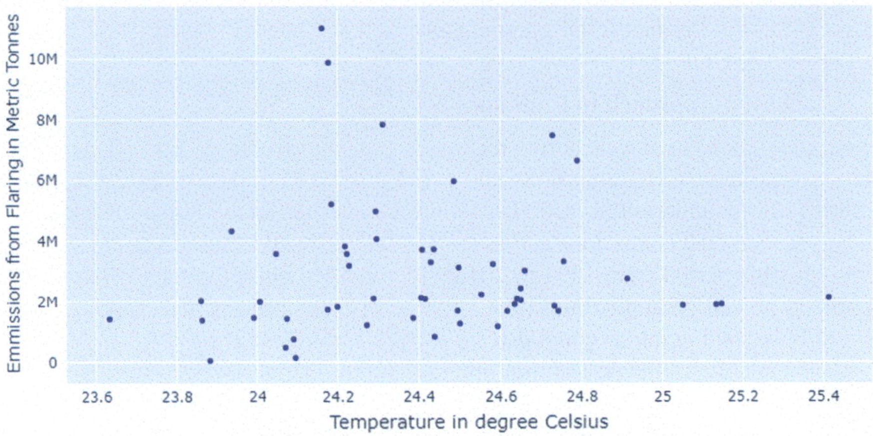

Fig. 5.3 Scatter plot of emissions from flaring versus temperature

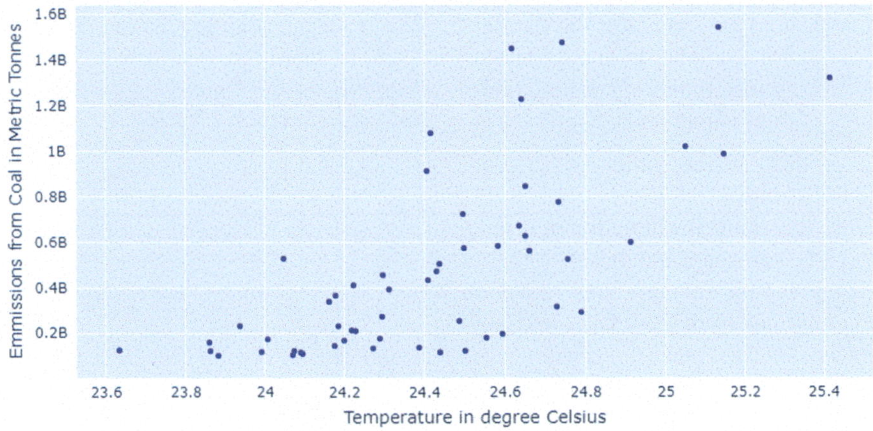

Fig. 5.4 Scatter plot of emissions from coal versus temperature

5.3 Building a Model

As discussed in the previous chapter, the dataset used has values from 1962 to 2016 containing six variables—"Year", "Oil", "Flaring", "Coal", "Gas", and "Temperature". In order to predict the annual temperature from 2021 onwards, the values for each of the variables have been predicted individually for 2017, 2018, 2019, and 2020 using ARIMA (autoregressive integrated moving average) modeling [2]. These forecasted values have been added to the existing dataset to extend the dataset up to the year 2020, with a total of 59 observations. Using this updated dataset, the future values of annual mean temperatures for 2021, 2022, 2023, 2024, and 2025 are predicted by applying ARIMAX (autoregressive integrated moving average exogenous) model

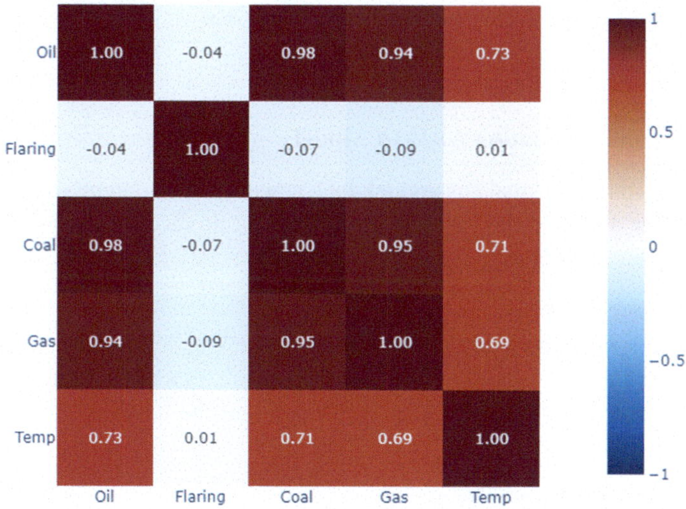

Fig. 5.5 Correlation heatmap between variables

[3]. In this model, "Oil", "Flaring", "Coal", and "Gas" have been considered as exogenous variables. There are a variety of forecasting models which could have been used, such as the VAR (vector autoregression) model. The issue with the VAR model is that it is based on numerous pre-defined assumptions. One of the assumptions which is most difficult to deal with is the interdependence of variables on each other, and the same is hard to satisfy by the present dataset. ARIMA model, on the other hand, makes no such assumption and is very easy to understand and implement.

5.3.1 Experimental Results

Python programming language is used to perform the analysis of the dataset prepared. The first step is to import the required libraries like pandas, numpy, and matplotlib, which are open-source libraries. Pandas library is a primary library for data analysis which wraps two libraries with exclusive tasks of data visualization (matplotlib) and numerical operations (numpy). The code to import the libraries is shown below:

```
import pandas as pd
import numpy as np
import matplotlib.pyplot as plt
%matplotlib inline

from statsmodels.tools.eval_measures import rmse, aic
from pmdarima import auto_arima
from statsmodels.tsa.statespace.sarimax import SARIMAX
from sklearn.metrics import mean_squared_error
from math import sqrt
```

```
from pandas.tseries.offsets import DateOffset

import warnings.
```

A. **Building and Evaluating ARIMA Model**

ARIMA is a statistical analysis model that stands for autoregressive integrated moving average and is applied to time series data for prediction based on historical data.

i. Determining the order of the ARIMA Model

Using the pmdarima library, we use auto_arima function to determine the best possible order for our variable. For this purpose, a function has been created, namely, det_order(a,var), which takes dataframe and the name of the variable as the arguments.

```
def det_order_ARIMA(a,var):
#Determining order
stepwise_fit = auto_arima(a[var],trace=True)

stepwise_fit.summary().
```

ii. Fitting the training model for ARIMA forecasting

In order to predict the values for the variables "Oil", "Flaring", "Coal", "Gas", and "Temperature" for 2017 to 2020, the dataset with 55 observations is split into two parts—training and testing. The training dataset has 51 observations, whereas the test dataset has four observations. The model has been built by using the training dataset to predict the values not included in the training dataset. This is achieved by importing the ARIMA function from statsmodels.tsa.arima_model and applying it to the dataset. A function named test_fit_ARIMA() has been created, which takes arguments as the number of testing observations (n). The same can be seen below:

```
def test_fit_ARIMA(n):

#Splitting Data into training and testing data
print(f"Shape of the dataset: {a.shape}\n")
global train
train = a.iloc[:-n]
global test
test = a.iloc[-n:]
print(f"Shape of the Training dataset: {train.shape}\nShape of the
Testing dataset: {test.shape}\n")

#Fitting training model
model = ARIMA(train[var],order=order)
model = model.fit()
print(model.summary)
```

iii. Evaluating the Model

The model is evaluated by comparing the predicted values with that of the values in the test dataset. Once the model correctly predicts the data, the next step is to fit the model on the entire dataset and predict values for the years 2017, 2018, 2019, and 2020. A function named eval_ARIMA() has been created to carry out this task.

```
def eval_ARIMA():
#Predicting based on training model
start = len(train)
end = len(train)+len(test)-1
global pred
pred = model.predict(start=start,end=end,typ='levels')
print(f"Values predicted on basis of Training dataset: \n{pred}\n")

#Predicted vs Testing Graph
fig1 = plt.figure(figsize=(15,8))
ax1  =  pred.plot(label  =  'Prediction  using  Training  Dataset',
legend=True,
figsize=(15,8))
ax2 = test[var].plot(label = 'Testing Dataset',legend=True)
fig1.add_subplot(ax1)
fig1.add_subplot(ax2)
fig1.show()

#Mean and RMSE
print(f'Mean: {test[var].mean()}\n')
rmse = sqrt(mean_squared_error(pred,test[var]))
print(f'RMSE: {rmse}\n')
```

Table 5.1 shows the best values for p, d, and q to be used for ARIMA model forecasting for each of the variables. For the purpose of model evaluation, it also includes the mean of the testing observations and Root Mean Squared Error (RMSE) obtained after fitting the model on the training dataset and the error percentage. It can be seen that the model correctly fits the data for all the variables with a relatively low percentage of error except for the variable "Flaring". This can be attributed to the fact that for "Flaring", India saw a drastic increase in CO_2 emissions released from "Flaring" in the year 1989. The model for "Flaring" does not fit the data very well as it does for the other variables.

The following graphs (Fig. 5.6, 5.7, 5.8, 5.9 and 5.10) are generated by calling the function for each variable. The graphs represent the observed values using the training

Table 5.1 Order selection of ARIMA model for forecasting variables for the years 2017, 2018, 2019, and 2020

Measure	Oil	Flaring	Coal	Gas	Temperature
Order selection (p, d, q)	(3, 2, 0)	(6, 0, 2)	(0, 2, 1)	(3, 1, 0)	(2, 1, 0)
Mean	542,727,904.5	1,838,411.25	1,445,070,980	110,162,456.75	24.98
RMSE	21,393,051.88	171,989.20	65,669,739.92	10,154,135.73	0.27
Error (%)	3.94	9.35	4.54	9.22	1.08

data and the predicted values using the test data for each variable. This helps to get an idea about how well the model fits the data. The blue line represents the predicted data, and the orange line represents the observed values for the corresponding years. It can be seen that both the lines are more or less overlapping for all the variables except for "Flaring". This shows that the values predicted by our model correctly resemble the actual values.

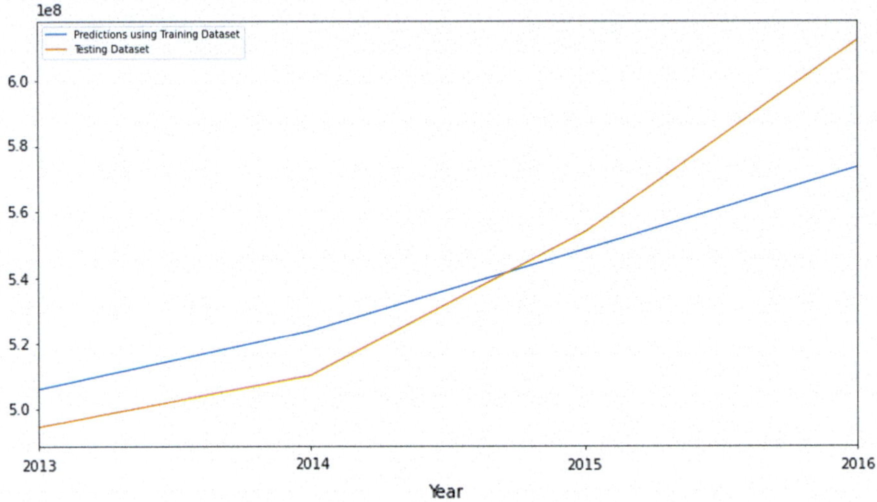

Fig. 5.6 Predicted versus actual data for emissions from "Oil"

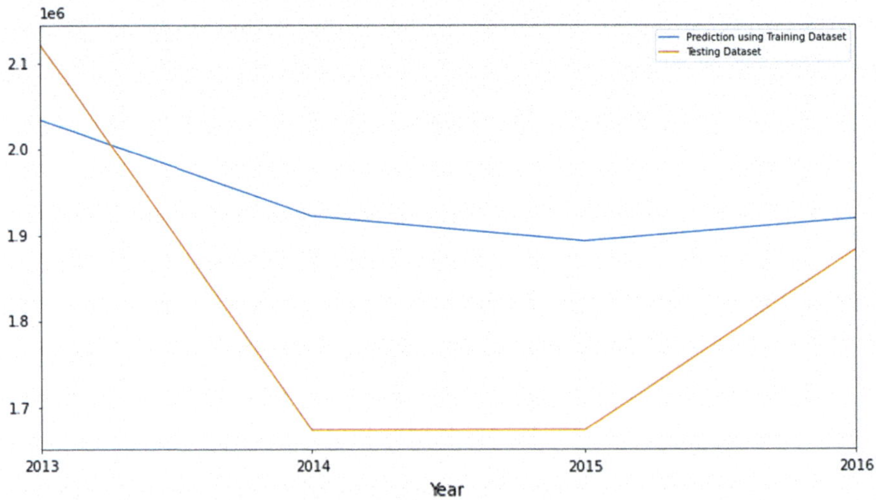

Fig. 5.7 Predicted versus actual data for emissions from "Flaring"

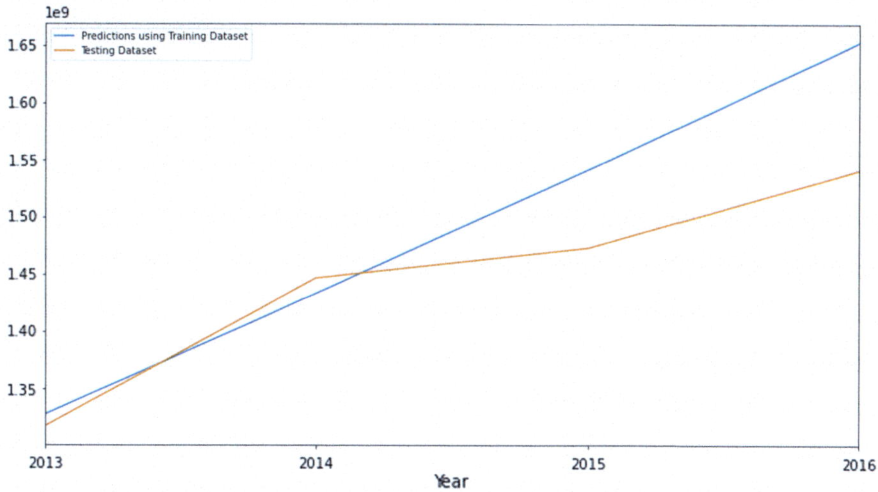

Fig. 5.8 Predicted versus actual data for emissions from "Coal"

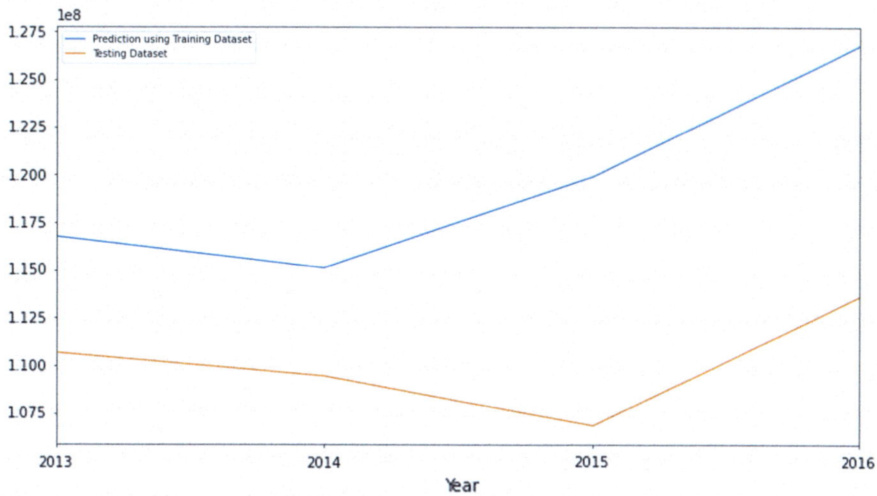

Fig. 5.9 Predicted versus actual data for emissions from "Gas"

iv. ARIMA forecasting for 2017, 2018, 2019, and 2020

In order to forecast the values for 2017, 2018, 2019, and 2020, the model is fit on the entire dataset, and the values are obtained. All these methods have been integrated into one function named test_fit (a, var, n, n1, order). This function takes the arguments—dataframe (a), variable (var), number of testing observations (n), the number of future predictions (n1), and the best order for the ARIMA model (order). The value of the argument n1 in the Python function is created as "the number of

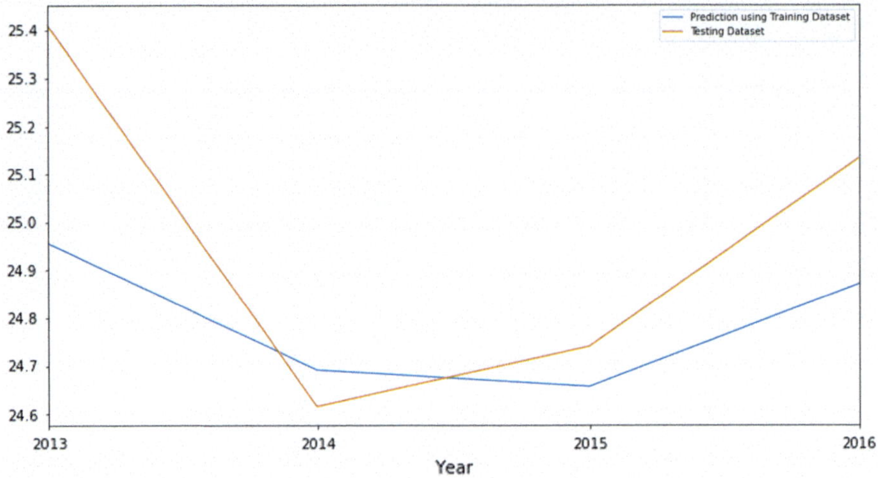

Fig. 5.10 Predicted versus actual data for "Temperature"

future predictions"—1. So, to make the predictions for the next 5 years, the value that is entered in the function is n−1, i.e., 5–1 = 4. The same can be seen in the code below:

```
def forecast ARIMA(a,var,n,n1,order)
#Fitting on entire dataset
model2 = ARIMA(a[var],order=order)
model2 = model2.fit()

#Predicting values for years 2017, 2018, 2019 and 2020
pred2 =
model2.predict(start=len(a),end=len(a)+n1,typ='levels').rename
('ARIMA Predictions') print(f"Predicted values: \n{pred2}\n")

#Plotting future values
fig2 = plt.figure(figsize=(15,8))
ax3 = pred2.plot(figsize=(15,8),legend=True)
fig2.add_subplot(ax3)
fig2.show()

#Forecasted Graph
pred2 = pred2.to_frame()
pred2.index.names=['Year']
ax1 = a.plot(y=var,label='Observed')
pred2.plot(y='ARIMA Predictions',label='Forecasted',ax=ax1,
figsize=(15,10))
```

The above-defined functions are called for each variable to get the ARIMA forecasted values for 2017, 2018, 2019, and 2020.

Table 5.2 shows the forecasted values for each variable after fitting the model to the entire dataset.

Table 5.2 ARIMA forecasted values for 2017, 2018, 2019, and 2020

Year	Oil	Flaring	Coal	Gas	Temperature
2017	641,648,024	2,055,813	1,619,816,616	115,863,676	24.87
2018	672,181,005	2,092,789	1,699,343,176	117,949,257	24.79
2019	719,831,730	2,048,963	1,780,015,416	118,259,570	25.05
2020	768,747,277	2,026,223	1,861,833,338	120,244,434	24.99

These forecasted values have been presented in the following graphs shown in Fig. 5.11, 5.12, 5.13, 5.14 and 5.15. While the blue line represents the observed values for each variable, the orange denotes the forecasted values for the years 2017, 2018, 2019, and 2020 using ARIMA forecasting.

B. **Building and Evaluating ARIMAX Model**

The ARIMAX model is simply an extension of the ARIMA model, which considers multivariate time services data with exogenous variables.

i. Determining the order of the ARIMAX Model

After obtaining the values for the years 2017, 2018, 2019, and 2020 for all the variables, the annual mean temperature is predicted using the ARIMAX model for the years 2021, 2022, 2023, 2024, and 2025 by taking "Oil", "Flaring", "Coal", and "Gas" as exogenous variables. The dataset has been updated by incorporating the ARIMA forecasted values for the years 2017, 2018, 2019, and 2020. It may be noted here that now the df dataset contains values up to the year 2020. The code snippet is provided below:

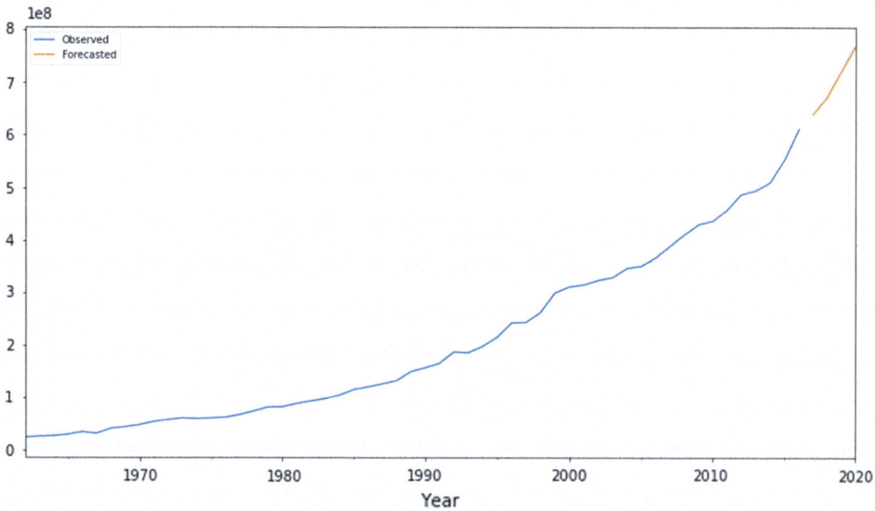

Fig. 5.11 Forecasted values of emissions from "Oil" for 2017, 2018, 2019, and 2020

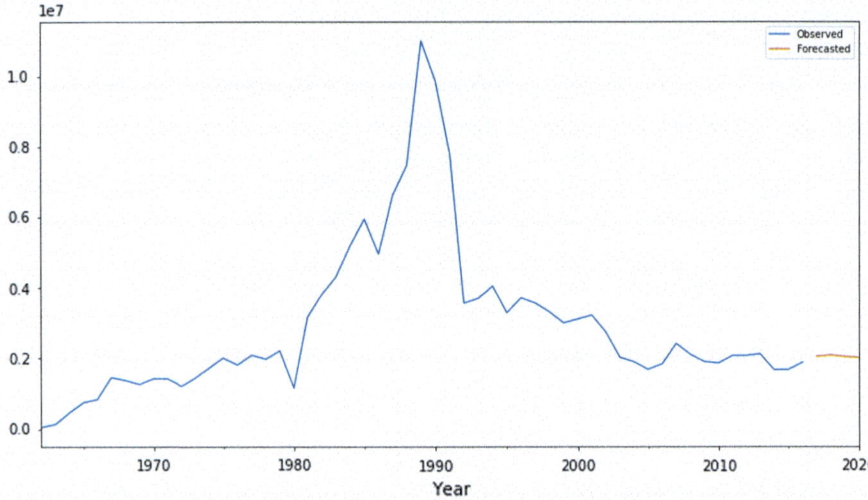

Fig. 5.12 Forecasted values of emissions from "Flaring" for 2017, 2018, 2019, and 2020

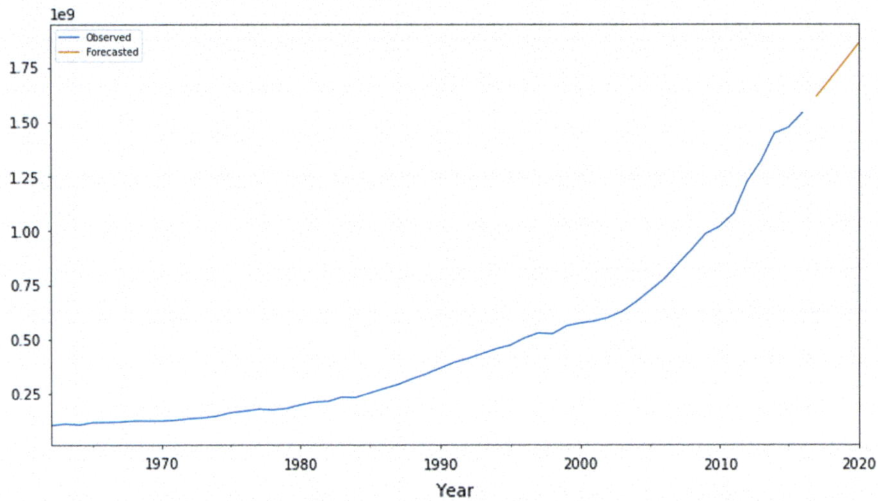

Fig. 5.13 Forecasted values of emissions from "Coal" for 2017, 2018, 2019, and 2020

```
df = pd.read_excel('Fuel wise 2020.xlsx',parse_dates=['Year'])
df.set_index('Year',inplace=True)
df.head()
```

After this, the best order for the model is determined with the help of following codes:

```
stepwise_fit = auto_arima(df['Temp'], trace=True)
stepwise_fit.summary()
```

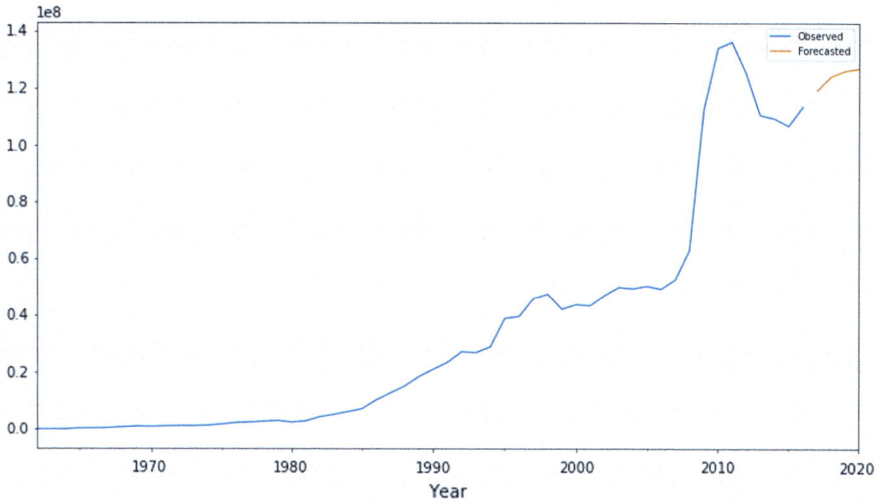

Fig. 5.14 Forecasted values of emissions from "Gas" for 2017, 2018, 2019, and 2020

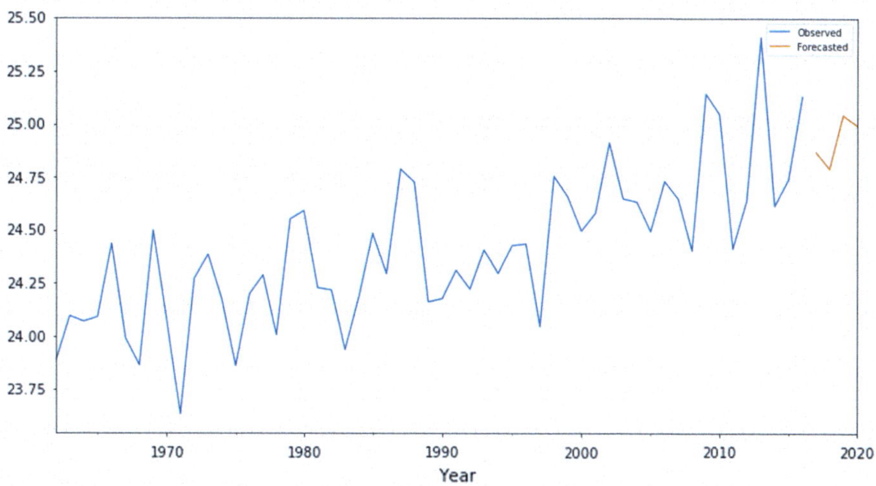

Fig. 5.15 Forecasted values of "Temperature" for 2017, 2018, 2019, and 2020

ii. Fitting the training model for ARIMAX forecasting

In case of ARIMAX modeling, the last 6 observations out of the 59 are considered as testing dataset and the rest of the observations are included in the training dataset. The following code is used to split the dataset into training and testing categories:

```
print(f'Shape of the dataset: {df.shape}\n')
train = df.iloc[:-6]
test = df.iloc[-6:]
```

```
print(f'Shape of the Training dataset: {train.shape}\nShape of the
Testing dataset: {test.shape}\n)

#Fitting training model
model = SARIMAX(train['Temp'],order = (2,1,1),exog =
train[['Oil','Flaring','Coal','Gas']])
model = model.fit()
model.summary()
```

iii. Evaluating the Model

The procedure to evaluate the model is similar to that of ARIMA forecasting mentioned above. Here, the SARIMAX function is imported from statsmodels.tsa.statespace.sarimax library. It may be noted that although the SARIMAX function is used, it becomes an ARIMAX model when seasonal_order argument is not passed in the function. As a next step to build the model, it is essential to ensure that the time series under study is stationary. For the current study, this requirement of stationarity has been achieved by using "trace = True" argument available in the auto_arima function. This ensures that the time series data is stationary before any modeling process is carried out. The evaluation of the model is done by executing the following code:

```
#Predicting based on training model
start = len(train)
end = len(train)+len(test)-1
pred =
model.predict(start=start,end=end,typ='levels',exog=test[['Oil',
'Flaring','Coal','Gas']]) print(pred)

#Predicted vs Testing Graph
pred.plot(label='Predicted',legend=True)
test['Temp'].plot(legend=True)

#Mean and RMSE
test['Temp'].mean()
rmse = sqrt(mean_squared_error(pred,test['Temp']))
print(rmse)
#Mean and RMSE
test['Temp'].mean()
rmse = sqrt(mean_squared_error(pred,test['Temp']))
print(rmse)
```

Table 5.3 shows the optimal order selection of p, d, and q for forecasting "Temperature" by applying the ARIMAX model. It also presents the mean and RMSE of the testing observations as well as the error percentage in order to get an idea of how the model performs. It can be seen that the model fits the data well, and the error percentage is relatively low.

Figure 5.16 depicts the predictions made based on the training data along with observed data. The blue line represents the predicted values, and the orange line indicates the observed values. The graph shows that the predicted values are close to the actual values, implying that the model correctly resembles the actual values.

Table 5.3 Order selection of ARIMAX model for forecasting "Temperature" for the period 2021–2025

Measure	Temperature
Order selection (p, d, q)	(2, 1, 1)
Mean	24.93
RMSE	0.22
Error (%)	0.88

Fig. 5.16 Predicted versus actual data for temperature by using ARIMAX forecasting

iv. ARIMAX Forecasting for 2021, 2022, 2023, 2024, and 2025

After confirming that the model fits the data correctly, the next step is to build the model on the entire dataset. The snippet of the code used is mentioned below:

```
#Fitting on entire dataset
model2 =
SARIMAX(df['Temp'],order=(2,1,1),exog=df[['Oil','Flaring',
'Coal','Gas']]) model2 = model2.fit()

#Predicting future values
pred1 =
model2.predict(start=len(df),end=len(df)+4,typ='levels',exog=df.
iloc[-5:][['Oil','Flaring','Coal','Gas']]).rename('ARIMAX Predic-
tions') print(pred1)

#Plotting future values
pred1.plot(figsize=(12,5),legend=True)

#Forecasted graph
pred1 = pred1.to_frame()
pred1.index.names=['Year']
ax = df.plot(y='Temp',label='Observed',)
```

Table 5.4 ARIMAX forecasted values for "Temperature" for the period 2021–2025

Year	Temperature
2021	24.78
2022	24.89
2023	24.88
2024	24.89
2025	25.02

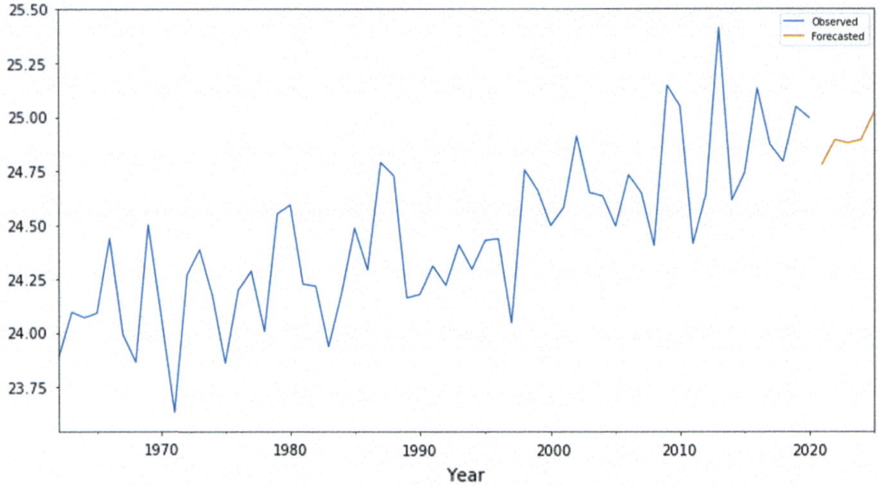

Fig. 5.17 Forecasted values of "Temperature" for 2021, 2022, 2023, 2024, and 2025

```
pred1.plot(y='ARIMAX         Predictions',label='Forecasted',ax=ax,
figsize=(15,10))
```

Table 5.4 shows the forecasted values of temperature for the years 2021, 2022, 2023, 2024, and 2025 based on the ARIMAX model. The forecasted values are also presented in Fig. 5.17. The blue line represents the actual values, whereas the orange line represents the values that have been forecasted using ARIMAX modeling.

It is visible from Fig. 5.17 that the trend of average land temperature has been increasing continuously, starting with 23.88 °C in 1962 to 24.18 °C in 1990, 24.74 °C in 2015, and 25.02 °C in 2025 (predicted value).

5.4 Conclusions

The ARIMA and ARIMAX model predictions reveal that if no drastic and immediate steps are implemented, emissions from "coal", "oil", "natural gas", and "flaring" will continue to rise in the near future at the current rate of consumption and production.

These findings indicate that the average land temperature in India is predicted to rise by 0.2 °C in the coming years.

"Coal", a non-renewable energy source, is by far the most significant contributor to energy-related emissions [4]. Coal's abundance, as well as the cheap and simple generation of electricity from it, makes it the most popular energy source not only in India but also around the world. However, due to overexploitation of the source, this abundance is quickly turning into scarcity. It not only depletes coal stocks at an alarming rate but also contributes considerably to emissions and, as a result, climate change. Despite the fact that coal is abundant in India, it cannot be replaced quickly enough to meet the needs of the Indian population. Oil and natural gas reserves in India are limited. There are just a few geographical locations where these reserves can be found. India will soon be on the verge of extinction of these reserves at its current rate of production and consumption [5].

As a result, India has a stringent and pressing need to transition to greener energy sources in order to meet its energy needs. India has the geographical advantage of being able to harness cleaner sources of energy to meet its need, just as it does with coal. The renewable energy resources of India and the energy produced for the year 2020 are shown in Table 5.5 [6]. It is clear that practically every state has the ability to harness energy from renewable sources. India has huge potential to create power and energy from wind, solar, hydro, and biomass plants due to its geographical location.

In order to control, if not reduce, the average land temperature, stringent policies can be introduced, systematic execution plans can be formulated, more standards can be implemented, and general awareness can be created towards climate conservation. This will help India to be on pace to meet the net-zero goal at the earliest.

To conclude, some additional measures are suggested below:

1. India has a large amount of potential for generating renewable energy that is yet to be fully realized. Therefore, the central government should collaborate with state governments to develop a blueprint for harnessing energy from greener sources in the future, based on each state's capabilities.
2. New rules and reforms are urgently needed to avoid the energy sector from over-exploiting non-renewable resources. It has also been discovered that a significant amount of electricity generated is squandered throughout the distribution process. To combat energy waste during distribution, adequate criteria, as well as fines and penalties, must be implemented.
3. An appropriate budget must be set aside for research and development. Research and development in this subject will eventually lead to more efficient technology that will assist in combating climate change more quickly.

The catastrophic effects of climate change should be taken as a responsibility not only by the government but also by the individuals. The goal of mitigating cannot be accomplished alone through the efforts of the government.

Table 5.5 State-wise overview of energy generation in India

Sl. No	States/UTs	Wind power@100 m	Small hydro power	Biomass power	Cogeneration-bagasse	Waste to energy*	Solar energy	Total	Distribution (%)
1	Andhra Pradesh	44,229	409.32	577.7	300	123	38,440	84,079	7.66
2	Arunachal Pradesh	–	2064.92	8.2	–	–	8650	10,723	0.98
3	Assam	–	201.99	211.8	–	8	13,760	14,182	1.29
4	Bihar	–	526.98	618.7	300	73	11,200	12,719	1.16
5	Chhattisgarh	77	1098.2	235.8	–	24	18,270	19,705	1.80
6	Goa	1	4.7	25.7	–	–	880	911	0.08
7	Gujarat	84,431	201.97	1220.7	350	112	35,770	122,086	11.12
8	Haryana	–	107.4	1332.6	350	24	4560	6374	0.58
9	Himachal Pradesh	–	3460.34	142.2	–	1.5	33,840	37,444	3.41
10	Jammu & Kashmir	–	1707.45	42.6	–	–	111,050	112,800	10.28
11	Jharkhand	–	227.96	90	–	10	18,180	18,508	1.69
12	Karnataka	55,857	3726.49	1130.5	450	–	24,700	85,864	7.82
13	Kerala	1700	647.15	1044.4	–	36	6110	9538	0.87
14	Madhya Pradesh	10,484	820.44	1364	–	78	61,660	74,406	6.78
15	Maharashtra	45,394	786.46	1887.3	1250	287	64,320	113,925	10.38
16	Manipur	–	99.95	13.4	–	2	10,630	10,745	0.98
17	Meghalaya	–	230.05	11.11	–	2	5860	6103	0.56
18	Mizoram	–	168.9	1	–	1.5	9090	9261	0.84
19	Nagaland	–	182.18	10.2	–	–	7290	7482	0.68
20	Odisha	3093	286.22	246.4	–	22	25,780	29,428	2.68

(continued)

Table 5.5 (continued)

Sl. No	States/UTs	Wind power@100 m	Small hydro power	Biomass power	Cogeneration-bagasse	Waste to energy*	Solar energy	Total	Distribution (%)
21	Punjab	–	578.28	3172.1	300	45	2810	6905	0.63
22	Rajasthan	18,770	51.67	1038.9	–	62	142,310	162,233	14.78
23	Sikkim	–	266.64	2.3	–	–	4940	5209	0.47
24	Tamil Nadu	33,800	604.46	1069.5	450	151	17,670	53,745	4.90
25	Telangana	4244	102.25	–	–	–	20,410	24,756	2.26
26	Tripura	–	46.86	2.9	–	1.5	2080	2131	0.19
27	Uttar Pradesh	–	460.75	1616.7	1250	176	22,830	26,333	2.40
28	Uttarakhand	–	1664.31	23.7	–	5	16,800	18,493	1.69
29	West Bengal	2	392.06	396	–	148	6260	7198	0.66
30	Andaman & Nicobar	8	7.27	–	–	–	–	15	0.00
31	Chandigarh	–	–	–	–	6	–	6	0.00
32	Dadar & Nagar Haveli	–	–	–	–	–	–	–	–
33	Daman & Diu	–	–	–	–	–	–	–	–
34	Delhi	–	–	–	–	131	2050	2181	0.20
35	Lakshadweep	8	–	–	–	–	–	8	0.00
36	Puducherry	153	–	–	–	2.5	–	156	0.01
37	Others*	–	–	–	–	1022	790	1812	0.17
All India Total		302,251	21,134	17,536	5000	2554	748,990	1,097,465	100.00
Distribution (%)		27.54	1.93	1.60	0.46	0.23	68.25	100.00	

References

1. Plotly Express in Python, Plotly (2022). Retrieved from https://plotly.com/python/plotly-express/
2. Hyndman RJ, Athanasopoulos G (2018) Forecasting: principles and practice, 2nd edn. OTexts, Melbourne, Australia. https://OTexts.com/fpp2
3. Hyndman RJ (2020) The ARIMAX Model Muddle. Retrieved from https://robjhyndman.com/hyndsight/arimax/
4. Generation of thermal power from raw coal. Ministry of Coal, Government of India (2022). Retrieved from https://coal.nic.in/en/major-statistics/generation-of-thermal-power-from-raw-coal
5. Mukharji A (2021) Why India is on the brink of an unprecedented power crisis. BBC News. Retrieved from https://www.bbc.com/news/business-58824804
6. Energy Statistics 2021 (2021) Ministry of Statistics and Program Implementation, Government of India. Retrieved from https://www.mospi.gov.in/documents/213904/957047//Energy%20Statistics%20India%202021162444418 6470.pdf/4012db31-691f-df9c-2cac-cd073efa1346

Chapter 6
Impact of Household Emissions on Climate Change in India—Literature Review and Data Preparation

6.1 Introduction

Climate change has always been a topic of discussion all over the world. But why? Is it a matter of concern? Climate change isn't a new phenomenon that the earth is going through. Our earth has gone through climate change naturally in the past too. Now, it is a matter of concern because of the speed at which the climate is changing. Climate change has caused many destructions of life on earth in the past. There were some massive changes in the structure of lands leading to the extinction of many species. But eventually, the earth finally adapts to the changing climate. However, the speed at which the climate is changing in the current times is substantial, and therefore it will be very difficult for life on earth to adapt to this rapid change, and hence, climate change is a matter of concern. There are predominantly two causes of climate change. First is climate change due to natural activities of the earth, such as volcanic eruptions. The second is anthropogenic causes, which means climate change occurs due to the contribution of human activities. In anthropogenic activities, climate change occurs due to the greenhouse effect originating from the Greenhouse Gases (GHGs). The GHG absorbs the energy from the sun and emits heat energy in all directions. When the balance of GHG in the atmosphere increases, heat energy gets emitted in the direction of the earth and also goes up, thereby warming the earth's atmosphere and leading to climate change [1].

The United Nations Framework Convention on Climate Change (UNFCCC) was formed in June 1992 to deal with climate change and reduce GHG emissions [2]. Limits are set for GHG emissions and targets to reduce the same by extending the UNFCCC with a protocol called the Kyoto Protocol in 1997 and getting a commitment from a few developed nations. With the Delhi Declaration (2002), Bali Roadmap (2007), and Copenhagen Accord (2009), commitments to reduce emissions were also undertaken by the developing countries [2]. The Paris Agreement, a legally binding international treaty signed by 196 nations in December 2015, aims to limit global warming to 2 °C while restricting the increase by 1.5 °C [2, 3]. According to the International Panel on Climate Change's (IPCC) special assessment on the effects

N. Sharma and P. K. De, *Towards Net-Zero Targets*, Advances in Sustainability Science and Technology, https://doi.org/10.1007/978-981-19-5244-9_6

of 1.5 °C, human activities have already caused global warming of 1 °C, and it will reach 1.5 °C between 2030 and 2052, emphasizing the significance of limiting global warming to 1.5 °C rather than 2 °C. [4]. Climate change issues are addressed by India through the launch of Initial National Communication to the UNFCCC, focusing on preserving the GHG concentrations in the atmosphere and preventing the dangerous interference of humans with the climate [5]. IPCC works in the preparation of reports that share knowledge about climate change, its impact, and future risks. The [8] guidelines give all the important and necessary information that anyone needs to know about GHGs. Information regarding GHG calculation from different sources such as energy, agriculture, industrial processes, waste, land use, and other sectors like residential, buildings, etc., is also provided. According to the guidelines by [8], carbon dioxide (CO_2), methane (CH_4), nitrous oxide (N_2O), and chlorofluorocarbons (CFCs) are among the gases to be held responsible for the greenhouse effect. However, CO_2 is the GHG which is widely studied and whose contribution to climate change is of vital importance [6].

Based on a study in India, total household energy consumption constitutes almost 75 percent of the total energy consumption [7]. GHG emissions from India's domestic/household/residential sector are primarily contributed by electricity usage in households and fuel consumption by households. Considering both the urban and rural areas, the majority of the fuels used for household purposes are LPG, kerosene, firewood, crop residues, and animal dung [8]. Despite having low per capita emissions, India is the largest CO_2 emitter after China and the United States. The emissions from Indian and Chinese cities are similar in magnitude, but those from Indian cities are much lower than those of US cities. India's CO_2 emissions increased by around 5 percent each year between 2010 and 2018, according to available estimates. Average monthly fossil fuel (coal, oil, natural gas, and carbonates) emissions of CO_2 stood at 190Mt (Million tons) for the year 2020 [9, 10]. From the start of the year 2020, as the COVID-19 pandemic struck the world, various countries started implementing lockdown policies to curb the coronavirus spread, which pretty much altered the patterns of energy demand worldwide, resulting in a significant reduction in CO_2 emission [11]. Global CO_2 emissions decreased up to 17 percent by April 2020 compared to emission levels in 2019 [11].

This chapter is divided into four sections. The next section (Sect. 6.2) discusses the trends in the household/residential sector, generalizing types of Household Carbon Emissions (HCE), establishes the resemblance between residential and primary energy household emissions and methodologies used in calculating HCE, and various literature related to HCE. Following that, Sect. 6.3 covers the materials and methods used for the problem statement. Finally, Sect. 6.4 concludes the chapter, followed by references.

6.2 Literature Review

The attempt in this section is to describe and compare the research findings related to the HCE. The study has been carried out using two approaches—systematic literature review and contemporary literature review. A systematic literature review is performed to show the relevance of the research towards household/residential carbon emissions. A contemporary literature review is carried out to understand the previous work and identify the research gaps, if any.

6.2.1 Systematic Literature Review

The motive behind adopting a systematic literature review for the household/residential sector is to identify the trends in this direction and emphasize the need to study this important topic. The SCOPUS database has been used to search for research works with the keywords "household CO_2 emissions", "household carbon emissions", "residential carbon emissions", and "residential CO_2 emissions". There are 163 studies registered in the database from 2008 to 2020. Figure 6.1 clearly shows that the number of studies in the above areas is steadily increasing year on year, with the maximum number of studies carried out in 2020, i.e., 30. The dotted line in the graph is the trendline which explains the increasing trend in research publications on household and residential CO_2 emissions.

It is evident from Fig. 6.2 that China is leading in the efforts put in publishing the research relating to household/residential carbon emissions, with 103 studies. This is followed by Australia (23), the United States (19), and Japan (17). At the bottom, only two studies are registered in the SCOPUS database as far as India is concerned. The dearth of research studies conducted in India clearly explains the need for the study.

Fig. 6.1 Year-on-year research publications (Documents) in SCOPUS

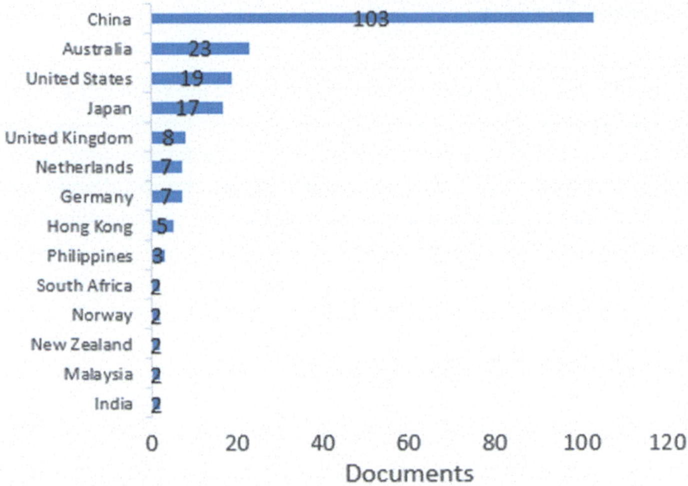

Fig. 6.2 Country-wise research publications (Documents) in SCOPUS

The Pie chart shown in Fig. 6.3 displays documents by type—"Article", "Conference Papers", "Book Chapter", "Review", "Editorial", and "Note". The "Article" type has the largest proportion of share (85 percent), followed by the "Conference Paper" (11 percent). The percentage share clearly indicates that the number of research-based publications is very less.

Lastly, the funnel chart presented in Fig. 6.4 depicts the publication by subject area such as "Environmental Science", "Earth and Planetary Science", "Computer Science", etc. The chart shows that "Environmental Science", "Energy", and "Engineering" are the areas where 65 percent of the studies are concentrated. "Environmental Science" (30 percent) is the largest area, whereas "Energy" and "Engineering" have 22 percent and 13 percent of the total studies, respectively. The "Decision Science" has a very negligible number of publications in the context of HCE and therefore presents enormous prospects for further study.

Fig. 6.3 Percentage share of publication type in SCOPUS

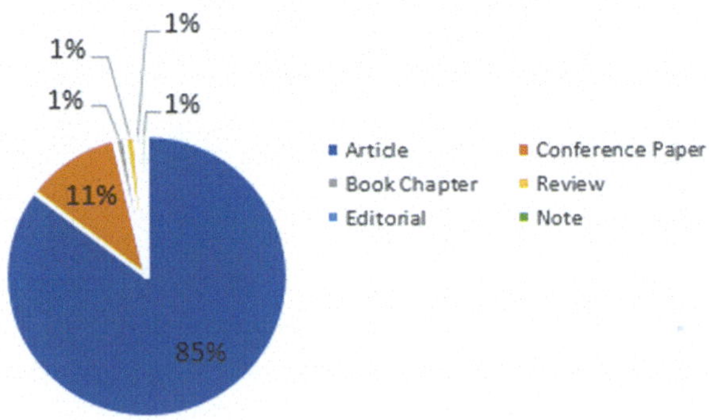

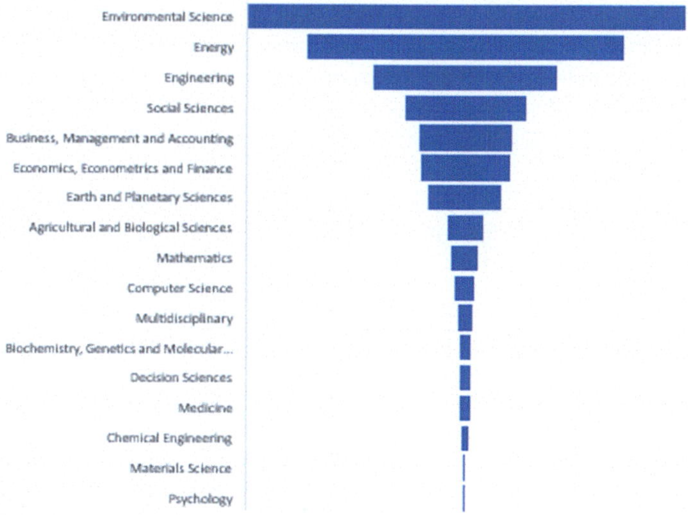

Fig. 6.4 Subject area-wise publication in SCOPUS

6.2.2 Contemporary Literature Review

This sub-section will provide a brief overview of the study carried out to:

- understand various types of HCE,
- study the similarity between residential emissions and primary energy household emissions,
- explore the methods for calculating HCE,
- refer to other related works, and
- study clustering approach to emissions.

A. **Understanding Various Types of HCE**

In general, HCE can be segregated into three categories—primary energy HCE, secondary energy HCE, and household consumption HCE. CO_2 emissions from the use of coal, gas, oil, and other fuels can be included in direct primary energy HCE. Secondary energy HCE includes indirect combustion of energy consumption such as electricity and heat. Food, clothes, housing, household equipment, medical care, transportation, communications, culture, education, entertainment, and other services are examples of indirect household energy consumption that can be included in the household consumption HCE [2].

B. **The Similarity between Residential Emissions and Primary Energy Household Emissions**

The [8] guidelines define the emissions from the residential sector as "Emissions from fuel combustion in households". Referring to the definition of "residential emissions" by the IPCC, the dissimilarity between household carbon emissions (HCE) and residential CO_2 emissions lies in the emissions from households due to the usage of fuels. All possible sources contributing to the emissions from households are taken into account in HCE. This problem arises due to the usage of different terminologies and the limitation of a globally standardized definition of HCE. Furthermore, there are also a variety of expressions which cause confusion [2]. Existing literature has focused on calculating the HCE using different methodologies and finding out important factors that affect the HCE [12]. Studies on HCE are usually confined to China, the United States, and the United Kingdom, while other countries have not received such attention [13].

C. **Methods used to calculate HCE**

In the literature, the authors have classified HCE into different types and calculated it in a variety of ways. [15] have briefly described the methodologies used to measure the HCE [12]. The authors have calculated direct emissions from primary energy HCE and secondary energy HCE by using the Emission Coefficient Method (ECM), whereas, for household consumption HCE (indirect emissions), methodologies like the Input–Output Method (IOM), Consumer Lifestyle Approach (CLA), and the Life-Cycle Analysis (LCA) are used.

D. **Review of works related to HCE**

Huang et al. [14] applied the ECM to calculate energy carbon emissions and CLA for product carbon emissions for the 30 provinces of China during 2006–2015. The authors have concluded that energy carbon emissions due to households are primarily from electricity and heat. In contrast, products HCE are due to consumption of high carbon emission categories like residences, food, transportation, and communication. Hence, the same results in a rapid growth trend in the HCE [11]. In addition, the authors have provided the extended STIRPAT (STochastic Impact by Regression on Population, Affluence and Technology) model of the influence factors and found out that population and per capita GDP are the driving factors contributing to the HCEs [16]. analyzed direct CO_2 emissions calculated by ECM and indirect CO_2 emissions based on IOM from urban and rural households in China's four megacities, namely, Beijing, Tianjin, Chongqing, and Shanghai, for the years 2007 and 2012. The authors have found that the total rural HCE is less than the total urban HCE. Also, Per capita Household CO_2 Emissions (PHCE) for rural cities are less than the urban cities for the four cities mentioned above, and electricity accounts for a larger proportion of direct HCE [14, 17] used the Multi-Regional Input–Output Table (MRIO) model and assessed the heterogeneity at the state level in the consumption-based accounting of CO_2 emissions in India. For the first time, estimation of the consumption-based inventories has been accounted at the state level by constructing a new Indian multi-state input–output table for 2015, based on the "Flegg Location Quotient Method". The authors have concluded that household consumption dominated 60–78 percent of the total consumption-based estimation at state the level. Moreover, the authors have partitioned the emissions into production-based and consumption-based emissions. As per the authors, the total consumption-based emissions in India are dominated by the consumption of households in affluent states [15, 18] used the ECM to calculate the direct HCE per capita and IOM for indirect HCE per capita in China for the years 2012, 2014, and 2016. As per their study, fuel type and dwelling type explain 70 percent of the direct HCE [16]. Besides, the LASSO model has been added to the existing literature for identifying the driving factors of the HCE [16].

Corinne le Quéré et al. [11] investigated per capita HCE from direct energy usage by the Indian cities for the years 2009–2010. Also, they analyzed the household emissions inventories with fixed effects and random effects panel regression models and concluded that the most important factors for household emissions are income and household size, followed by other factors like urban and dwelling unit characteristics, gender, and education level [13]. The consumption-based household carbon footprints of 623 districts in India have been calculated based on MRIO and compared by [19], explaining the variation by cultural, economic, and demographic factors [17]. Ramchandra et al. (2014) analyze and estimate GHGs from sectors like energy, domestic, transportation, industrial, livestock management, agriculture, and waste from major cities in India, namely, Delhi, Mumbai, Chennai, and Bangalore, using the equations given in the GHG inventories of the IPCC 1996 and 2006 guidelines. Besides, the authors have expressed the total CO_2 equivalent (CO_2 eq) in terms of Gross Domestic Product (GDP) and concluded that Chennai is the highest emitter of CO_2 eq per GDP [18].

E. **Clustering Approach to Emissions**

Anna et al. (2016) used a clustering technique to categorize the 28 European nations according to their GHG emissions and per capita GHG emissions in 2012 into 4 groups [19]. Ghanbari et al. (2020) conducted a cluster analysis using 18 factors, including total GHG emissions across the Middle East and Central Asia, by considering the time period from 1994 to 2014 and classified into 5 primary groups [20]. Mohan et al. (2018) applied the IPCC standards for measuring and tracking GHG data in India, taking into account diverse sectors and states [21].

Table 6.1 collates the description of the literature that is reviewed from different perspectives, such as—the area of the study, the country in which the research is carried out, the source of the data, the time period in which the data has been collected, and the different methods used to analyze and calculate the emissions.

Most of the available literature focuses on calculating, assessing, and identifying relevant components of household emissions across the globe. A clustering approach to household emissions for the dataset from Europe, the Middle East, and Central Asia is also available in some literature. However, there is a dearth of literature, which classifies Indian states based on household emissions. Therefore, the primary focus of this chapter and Chap. 7 are as mentioned below:

1. To find out the similarity between residential and primary energy household emissions.
2. To determine the direction of primary energy household emissions in India from LPG, kerosene, and diesel.
3. To classify Indian states based on their varying primary energy household emissions from CO_2, CH_4, and N_2O.

Table 6.1 Review of papers

Research paper	Year	Area of study	Country	Data period	Data source	Methods
Direct and Indirect Energy Requirements of Households in India [7]	2002	Household energy requirement	India	1983–84 and 1993–94	Commercial: Coal, PNG, Power Statistics. Various issues on "Energy" by CMEI, Electricity Statistics by CEA, TERI online Energy Database Non-Commercial: Regional Wood Energy Development Programme of the FAO (1997) Input–Output Tables for India	Input–Output Table
Comparative Analysis of Greenhouse Gas Emissions from Major Cities of India [18]	2014	(Domestic and other Sectors) Carbon Emissions	Delhi, Greater Mumbai, Chennai, Greater Bangalore	Activity Data—2009, 2010 Census Data—2003, 2007	Consumption Data: Government Reports of Central and State Organization, Census Data	Equations from IPCC Guidelines
CO_2 Emissions from Direct Energy use of Urban Households in India [9]	2015	HCE	India	2010	NSSO Consumer Expenditure Survey 2010	Simple linear regression (Fixed effect and random effect model & OLS), emissions = F1*electricity + F2* cooking fuels + F3*private transportation

(continued)

Table 6.1 (continued)

Research paper	Year	Area of study	Country	Data period	Data source	Methods
Household Carbon Emission Research: An Analytical Review of Measurement, Influencing Factors and Mitigation Prospects [13]	2015	Review on HCE	Global	–	–	–
Research of Varying Levels of Greenhouse Gas Emissions in European Countries Using the K-Means Method [19]	2016	Clustering Approach	Europe	2012	EUROSTAT	K-Means Clustering
The Scale Structure and Influencing Factors of Total Carbon Emissions from households in 30 provinces of China—Based on the extended STIRPAT Model [12]	2018	HCE	30 provinces of China	2006–2015	Regional Energy Balance Table: China's Statistical Year Handbook (2006–2015)	Energy Carbon Emission: ECM Product Carbon emission: CLA Influencing Factors: IPAT/STIRPAT Model

(continued)

Table 6.1 (continued)

Research paper	Year	Area of study	Country	Data period	Data source	Methods
Comparing Urban and Rural Household CO_2 Emissions—Case from China's Four Megacities: Beijing, Tianjin, Shanghai and Chongqing [14]	2018	HCE	Beijing, Tianjin, Shanghai and Chongqing	Direct—2007–2015, Indirect—2007 and 2012	Energy consumption Data: China Energy Statistical Handbook Direct CO_2 Coeff: IPCC Report Indirect CO_2 Emissions: China's Multi-Regional Input—Output Table 2007, 2012 and also http://www.ceads.net/, Indirect Emission data are given in the table Population: City's Statistical Handbook	Direct Carbon Emission: ECM, Indirect Carbon Emission: IOM
Household CO_2 Emissions: Current Status and Future Perspectives [2]	2020	Review on HCE	Global	–	–	–
Prioritizing Driving Factors of Household Carbon Emissions: An Application of the LASSO Model with Survey Data [16]	2020	HCE	China	2012, 2014 and 2016	Samples of China households from China Family Panel Studies (CFPS) Input–Output tables from WIOD Carbon emissions for China's 35 sectors from WIOD	Direct HCE: ECM Indirect HCE: IOM Influencing Factors: LASSO model

(continued)

Table 6.1 (continued)

Research paper	Year	Area of study	Country	Data period	Data source	Methods
The Scale and Drivers of Carbon Footprints in Households, Cities and Regions Across India [17]	2020	Household Carbon Footprint	India	2015, 2017, 2020	EORA Global Supply Chain Database, MRIO	(1) CF = Emission intensity (CO_2/$) *Leontief Inverse*Consumption expenditure
Urban and Rural Contribution to the GHG Emissions in the MECA Countries [20]	2020	Clustering Approach	The Middle East and Central Asia	1994–2014	World Bank Database, FAO, WUP, EDGAR, CDIAC, IEA	Hierarchical Clustering, Correlation Analysis, and Regression
Heterogeneity of Consumption-based Carbon Emissions and Driving Forces in the Indian States [15]	2021	Consumption-based and Production-based carbon emissions	India	2015	Production-based Carbon Emissions: GHG Platform of India (GHGPI), Consumption-based Emissions: Indian Supply-Use Table (SUT), CEIC datasets, ASI, MoSPI	Multi-regional Input–Output Model, "Flegg Location Quotient Method"

4. To identify important fuel sources and determine the factors that can lead to the cluster formation.
5. To recommend the ways to limit primary energy household emissions in India

6.3 Materials and Methods

This section presents the steps towards data preparation and discussions around probable algorithms to build a clustering model.

6.3.1 Data Preparation

Data preparation is the process of making a dataset ready for exploration and analysis purposes. The process involves multiple steps like—identifying the data sources, acquiring relevant data, cleaning, and wrangling the dataset to make it appropriate and usable.

A. **Data source**

The dataset used is from Phase III emissions estimates of the energy sector from the GHG Platform India (GHGPI), titled "Phase III (2005–2015) Emission Estimates Energy" [21, 22]. The GHGPI estimates the emissions as per the guidelines issued by [8] for the national GHG inventories. The data is maintained in multiple sheets in the excel file that records the following information:

a. Introduction of GHGPI website,
b. Estimation of the emissions, and
c. Fuel consumption data for different sources.

For this research, the "GHG Estimation" sheet has been used, as it contains the final estimation of the emissions data. The GHG estimation from the energy sector consists of the fuel combustion emissions from the production of energy, transportation sector, generation of electricity, and others, including the residential, agricultural, commercial, and fishing sectors, for the period between 2005 and 2015. The projected population data for the profiling is obtained from the Census of India [23].

It may be noted here that there is a dearth of data related to the residential sector emissions. The data from the GHG Platform India is used because it contains state-wise emissions data for the Indian residential sector emissions. The dataset is the time series data with the emissions in tons from 2005 to 2015.

B. **Data Cleaning for Analysis**

The study includes the "GHG Estimation" dataset extracted from GHGPI. The original dataset consists of 23 columns and 11,489 rows. Before using the dataset for Exploratory Data Analysis (EDA), the data is cleaned, aggregated, and grouped as per the requirement of the study. The columns in the original dataset are distributed in seven levels (Level 1–Level 7), covering sector-wise emissions. The columns under Level 5 and Level 7 are dropped in the absence of data values. Level 1 and Level 2 refer to the emissions from energy and fuel combustion, respectively, hence are not considered in the current study. Level 3 and Level 6 are renamed as "Sector" and "Fuel Type", respectively, for better understanding. The data in the "Sector" column is filtered to segregate "residential sector emissions" and further renamed as "primary energy household emissions". Columns "Gas" and "States" are considered as it is, and the rows containing NA (Not Available) values have been dropped. The columns for the year 2005–2015 have been selected and considered for the data exploration part. Table 6.2 displays the dataset that is used for the analysis.

The dataset is further processed for the clustering purpose of segmenting the Indian states based on household emissions. The dataset presented in Table 6.2 is transformed to present state-wise, fuel-type-wise gas emissions. The LPG, kerosene,

Table 6.2 Structure of the primary household emissions dataset used for EDA

Columns	Description
Sector	Only primary energy household emissions are considered
Fuel Type	LPG, Kerosene, Charcoal, Coal, Coke, Natural Gas, and Fuelwood
Gas	CO_2, CH_4, and N_2O
State	Only 35 Indian States and Union Territories
2005	Sector-Wise, Fuel-type-wise, Gas-wise, State-wise Emissions value for the year 2005
2006	Sector-Wise, Fuel-type-wise, Gas-wise, State-wise Emissions value for the year 2006
2007	Sector-Wise, Fuel-type-wise, Gas-wise, State-wise Emissions value for the year 2007
2008	Sector-Wise, Fuel-type-wise, Gas-wise, State-wise Emissions value for the year 2008
2009	Sector-Wise, Fuel-type-wise, Gas-wise, State-wise Emissions value for the year 2009
2010	Sector-Wise, Fuel-type-wise, Gas-wise, State-wise Emissions value for the year 2010
2011	Sector-Wise, Fuel-type-wise, Gas-wise, State-wise Emissions value for the year 2011
2012	Sector-Wise, Fuel-type-wise, Gas-wise, State-wise Emissions value for the year 2012
2013	Sector-Wise, Fuel-type-wise, Gas-wise, State-wise Emissions value for the year 2013
2014	Sector-Wise, Fuel-type-wise, Gas-wise, State-wise Emissions value for the year 2014
2015	Sector-Wise, Fuel-type-wise, Gas-wise, State-wise Emissions value for the year 2015

Table 6.3 Description of the state-wise household emission data

Columns	Description
States	35 Indian States and Union Territories
LPG_CO_2	Primary energy HCE from LPG in 1000 tons
Kerosene_CO_2	Primary energy HCE from Kerosene in tons
Diesel_CO_2	Primary energy HCE from Diesel in tons
LPG_CH_4	Primary energy CH_4 emissions from LPG in tons
Kerosene_CH_4	Primary energy CH_4 emissions from Kerosene in tons
Diesel_CH_4	Primary energy CH_4 emissions from Diesel in tons
Fuelwood_CH_4	Primary energy CH_4 emissions from Fuelwood in tons
LPG_N_2O	Primary energy N_2O emissions from LPG in tons
Kerosene_N_2O	Primary energy N_2O emissions from Kerosene in tons
Diesel_N_2O	Primary energy N_2O emissions from Diesel in tons
Fuelwood_N_2O	Primary energy N_2O emissions from Fuelwood in tons

diesel, and fuelwood are considered as "Fuel-Type", whereas CO_2, CH_4, and N_2O are considered as "Gas" emissions. As per the availability of the primary energy household emissions data for the states, gas, and fuel types, 35 Indian states and union territories are considered as objects in the data frame. The final dataset is in wide format with 11 attributes, as shown in Table 6.3. The dataset includes the emissions for the latest available year, i.e., 2015, for the purpose of clustering.

6.3.2 Selected Algorithms

With no prior knowledge about the expected shape of the clusters being formed, the hierarchical clustering and k-means clustering approaches are used. The hierarchical clustering generates a dendrogram that helps visualize the possible combination of clusters. The k-means algorithm creates the interpretable clusters by calculating the centroid and readjusting it in an iterative manner to fit the observations with precision.

A. **Agglomerative Hierarchical Clustering**

The principle behind hierarchical agglomerative clustering is to start with each cluster containing only one record and gradually combine the two closest clusters until only one cluster remains at the end, which contains all of the information. The closet records are determined by distance metric (e.g., Euclidean), whereas closet clusters are determined by linkage method (e.g., average group linkage). Agglomerative hierarchical clustering produces a dendrogram, which is useful in visualizing clusters. The steps followed to create agglomerative hierarchical clustering are mentioned below:

1. Begin by creating n clusters (each observation equals a cluster).
2. The two observations that are closest to each other are merged into a single cluster.
3. The two groups with the shortest distance are combined at each stage. This means that a single observation is either added to existing clusters or two clusters are joined.

B. **K-means Clustering**

K-means is a non-hierarchical strategy to construct a perfect cluster that involves pre-selecting the desired number of clusters k and assigning each data point to one of the k clusters in order to minimize a measure of cluster dispersion. In other words, the purpose is to divide the sample into a preset number k of non-overlapping clusters, each of which should be as homogeneous as possible in terms of the measurements used. The sum of squared Euclidean distances of records from their cluster centroid is one measure of within-cluster dispersion. The problem can be set up as an integer programming optimization problem. However, solving integer programs with a large number of variables is time-consuming. Therefore, clusters are frequently generated using a rapid, heuristic method that provides good (but not necessarily optimal) answers. The first step in the k-means method is to divide the cases into k clusters. Following that, the partition is changed to reduce the sum of each record's distances from its cluster centroid. The adjustment consists of allocating each record to the k centroids of the previous division that is closest to it. This results in a new division with a lesser total of distances than the previous one. The new cluster means are calculated, and the improvement step is repeated until the same is negligible. The steps for performing k-means clustering are mentioned below:

1. Begin by creating k initial clusters (user chooses k).
2. Each record is transferred to the cluster with the "closest" centroid at each stage.
3. Recalculate the centroids of clusters that have lost or acquired a record, and then repeat steps 2 and 3.
4. Stop shifting records between clusters if doing so causes cluster dispersion.

6.4 Conclusions

The chapter started with an introduction to climate change, discussed global initiatives taken to tackle it, and highlighted the household sector emissions in India. The literature review clearly shows that there is an increasing trend in research work around residential and household emissions, and the study is predominantly carried out in China, followed by Australia and the United States. The literature also established a resemblance between residential and primary energy household emissions by reviewing various works related to HCE. Some researchers have used a clustering approach to classify region-wise household emissions. However, very little research has been carried out for India, providing an opportunity for this research. Lastly, the identification of the data source, dataset preparation, and the chosen algorithms are presented to model a solution towards the net-zero target. The next chapter covers the exploratory data analysis to explain the structure and behaviors of data, followed by the application of unsupervised machine learning algorithms to classify Indian states based on their varying primary energy household emissions.

References

1. An overview from the Royal Society and the US National Academy of Sciences, Update 2020, The Royal Society, National Academy of Sciences https://royalsociety.org//media/Royal_Society_Content/policy/projects/climate-evidence-causes/climate-change-evidence-causes.pdf
2. Liu L, Qu J, Maraseni TN, Niu Y, Zeng J, Zhang L, Xu L (2020) Household CO_2 emissions: current status and future perspectives. Intern J Environ Res Public Health (19):7077. https://doi.org/10.3390/ijerph17197077
3. The United Nations Framework on Climate Change, What is Kyoto Protocol?. https://unfccc.int/kyoto_protocol
4. IPCC (2018) Global Warming of 1.5 °C. An IPCC Special Report on the impacts of global warming of 1.5°C above pre-industrial levels and related global greenhouse gas emission pathways, in the context of strengthening the global response to the threat of climate change, sustainable development, and efforts to eradicate poverty [Masson-Delmotte V, P Zhai, H-O Pörtner, D Roberts, J Skea, PR Shukla, A Pirani, W Moufouma-Okia, C Péan, R Pidcock, S Connors, JBR Matthews, Y Chen, X Zhou, MI Gomis, E Lonnoy, T Maycock, M Tignor, T Waterfield (eds)]. In Press https://www.ipcc.ch/site/assets/uploads/sites/2/2019/06/SR15_Full_Report_High_Res.pdf
5. Ministry of Environment and Forests GoI. India's Initial National Communication to the United Nations Framework Convention on Climate Change (2004). https://unfccc.int/resource/docs/natc/indnc1.pdf
6. IPCC (2006) Guidelines for National Greenhouse Gas Inventories. https://www.ipcc-nggip.iges.or.jp/public/2006gl/
7. Pachauri S, Spreng D (2002) Direct and indirect energy requirements of households in India. Energy Policy 30(6):511–523. https://doi.org/10.1016/s0301-4215(01)00119-7
8. Ministry of environment and Forests government of India, India: Greenhouse gases emissions (2007). https://www.iitr.ac.in/wfw/web_ua_water_for_welfare/water/WRDM/MOEF_India_GHG_Emis_2010.pdf
9. Ahmad S, Baiocchi G, Creutzig F (2015) CO2 emissions from direct energy use of urban households in India. Environ Sci Technol 49(19):11312–11320. https://doi.org/10.1021/es505814g
10. Andrew RM (2020) Timely estimates of India's annual and monthly fossil CO_2 emissions. Earth Syst Sci Data 12(4):2411–2421. https://doi.org/10.5194/essd-12-2411-2020
11. Corinne le Quéré C, Jackson RB, Jones MW, Smith AJP, Abernethy S, Andrew RM, De-Gol AJ, Willis DR, Shan Y, Canadell JG, Friedlingstein P, Creutzig F, Peters GP (2020) Temporary reduction in daily global CO_2 emissions during the COVID-19 forced confinement. Nat Clim Change 10(7):647–653. https://www.nature.com/articles/s41558-020-0797-x

12. Wang Y, Yang G, Dong Y, Cheng Y, Shang P (2018) The scale, structure and influencing factors of total carbon emissions from households in 30 provinces of China—based on the extended STIRPAT model. Energies 11(5):1125. https://doi.org/10.3390/en11051125
13. Zhang X, Luo L, Skitmore M (2015) Household carbon emission research: an analytical review of measurement, influencing factors and mitigation prospects. J Clean Prod 103:873–883. https://doi.org/10.1016/j.jclepro.2015.04.024
14. Huang R, Zhang S, Liu C (2018) Comparing urban and rural household CO_2 emissions—case from China's four megacities: Beijing, Tianjin, Shanghai, and Chongqing. Energies 11(5):1257. https://doi.org/10.3390/en11051257
15. Huang Q, Zheng H, Li J, Meng J, Liu Y, Wang Z, Zhang N, Li Y, Guan D (2021) Heterogeneity of consumption-based carbon emissions and driving forces in Indian states. Adv Appl Energy 100039. https://doi.org/10.1016/j.adapen.2021.100039
16. Shi X, Wang K, Cheong TS, Zhang H (2020) Prioritizing driving factors of household carbon emissions: an application of the LASSO model with survey data. Energy Econ 92:104942. https://doi.org/10.1016/j.eneco.2020.104942
17. Lee J, Taherzadeh O, Kanemoto K (2021) The scale and drivers of carbon footprints in households, cities and regions across India. Glob Environ Chang 66:102205. https://doi.org/10.1016/j.gloenvcha.2020.102205
18. Ramachandra TV, Hari Sridevi (2014) Comparative analysis of greenhouse gas emissions from major cities of India.02. https://www.researchgate.net/publication/297716139_Comparative_analysis_of_greenhouse_gas_emissions_from_major_cities_of_India
19. Kijewska A, Bluszcz A (2016) Research of varying levels of greenhouse gas emissions in European countries using the k-means method. Atmosp Poll Res 7(5):935–944. https://doi.org/10.1016/j.apr.2016.05.010
20. Ghanbari S, Mansouri Daneshvar MR (2021) Urban and rural contribution to the GHG emissions in the MECA countries. Environ Dev Sustain 23:6418–6452. https://doi.org/10.1007/s10668-020-00879-8
21. Mohan RR, Dharmala N, Ananthakumar MR, Kumar P, Bose A (2019) Greenhouse gas emission estimates from the energy sector in India at the sub-national level (Version/edition 2.0). New Delhi. GHG Platform India Report. http://www.ghgplatform-india.org/data-and-emissions/energy/GHGPI-PhaseIII-Methodology%20Note-Energy-Sep%202019.pdf
22. GHG Platform India. http://www.ghgplatform-india.org/energy-sector
23. Government of India, Census of India. https://censusindia.gov.in/Census_Data_2001/Projected_Population/Projected_Population.pdf

Chapter 7
Use of Unsupervised Machine Learning Algorithms to Segment Indian States Based on Primary Energy Household Emissions

7.1 Introduction

In Chap. 6, climate change and various initiatives to tackle it have been briefed and the primary energy household emissions have also been investigated by evaluating several published literature. The prior research works have been studied from different perspectives to understand the various areas of the research, as mentioned below:

a. Household Carbon Emissions (HCE).
b. Similarity between residential and primary energy household emissions.
c. Methods for calculating HCE.
d. Clustering approach to understand the severity of emissions.

Chapter 6 achieved the objective of comparing residential and primary energy household emissions to find the similarity between the two. In this chapter, the focus is to determine the direction of primary energy household emissions in India with the help of exploratory data analysis, which presents the trend and the variation of emissions across the Indian States.

Further, the chapter discusses the application of clustering algorithm based on CO_2 emissions for 35 States and Union Territories. In addition, profiling of the clusters is carried out by using CO_2 emissions and the population of these States. Finally, the recommendations are made towards limiting household carbon emissions.

The chapter is structured in four sections. The next section talks about exploratory data analysis (EDA) and presents the results from various perspectives. Section 7.3 covers the cluster analysis and profiling. The conclusions of the chapter are presented in Sect. 7.4.

N. Sharma and P. K. De, *Towards Net-Zero Targets*, Advances in Sustainability Science and Technology, https://doi.org/10.1007/978-981-19-5244-9_7

7.2 Exploratory Data Analysis (EDA)

Data preparation is the process of preparing the raw data into a clean dataset that can be used for further analysis. Chapter 6 has presented the sources of data, followed by a complete process of data preparation [1]. Once the dataset is ready, exploratory data analysis is performed to understand the structure and behavior of the data, and the same is depicted using various visualization tools.

The line chart and stacked bar charts are used in this section to explore the data. EDA is performed using excel to gain insights from the data, observe the direction, and understand the variation in emissions across different States and Union Territories in India.

7.2.1 CO_2 Emissions by Commonly Used Fuel Type

An attempt has been made to explore the level of emissions by fuel type (LPG, kerosene, and diesel), which is majorly used in the households of various States of India for the year 2015. Figure 7.1 clearly shows that Uttar Pradesh, Maharashtra, Tamil Nadu, West Bengal, and Karnataka are the top five major contributors of primary energy HCE, whereas Lakshadweep is the smallest emitter. Uttar Pradesh has the highest level of emissions from all the fuel types, i.e., LPG, kerosene, and diesel, solely because of the area and population of the State. The maximum and minimum CO_2 emissions from households are recorded at 10,879,412 tons and 3892 tons for Uttar Pradesh and Lakshadweep, respectively. Considering all the States, the average primary energy HCE (LPG, kerosene, and diesel) is 2064994 tons.

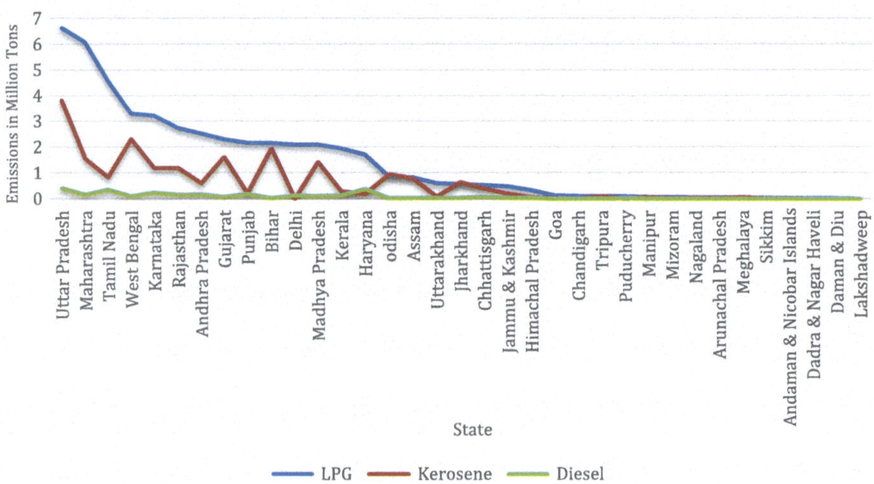

Fig. 7.1 Primary energy household emissions by fuel type and States in 2015

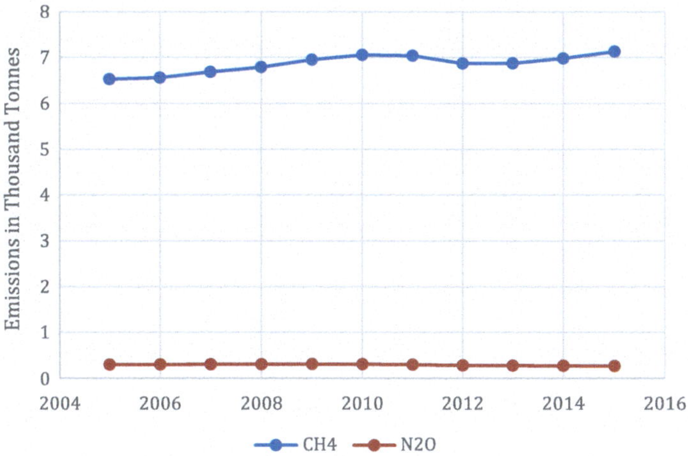

Fig. 7.2 Total primary energy household emissions of LPG, kerosene, and diesel by gas type

7.2.2 CH_4 and N_2O Emissions by Commonly Used Fuel Type

Based on the availability of emissions data for the States, Fig. 7.2 represents total primary energy household methane (CH_4) and nitrous oxide (N_2O) emissions from LPG, kerosene, and diesel for those States over the time period 2005–2015. It is clearly visible that CH_4 emissions increased from 6531 tons in 2005 to 7058 tons in 2010. A slight decline to 6874 tons is observed in 2012, followed by an increase to 7133 tons in 2015. Overall, a slight rise is seen in CH_4 emissions. However, the total primary energy household N_2O emissions from LPG, kerosene, and diesel remained constant throughout the period 2005–2015.

7.2.3 Emissions by All Fuel Types

In the sub-section 7.2.1, emissions from only the three most commonly used fuel types (LPG, kerosene, and diesel) have been analyzed for the year 2015—the latest data available in the dataset. However, in this sub-section, emissions from all the fuel types (LPG, kerosene, diesel, charcoal, coal, coke, and natural gas) have been considered. Figs. 7.3 and 7.4 present the average primary energy HCE from fuel usage from 2005 to 2010 and from 2011 to 2015, respectively, to represent the information with better clarity.

Figure 7.3 highlights that in most of the States, kerosene is a preferred fuel type as it was easily available at a lesser price during 2005–2010. Therefore, the primary energy HCE from the kerosene seem higher compared to other fuel types.

It is observed from Fig. 7.4 that during the period from 2011 to 2015, LPG was the most consumed fuel type in the majority of the States. This may be because

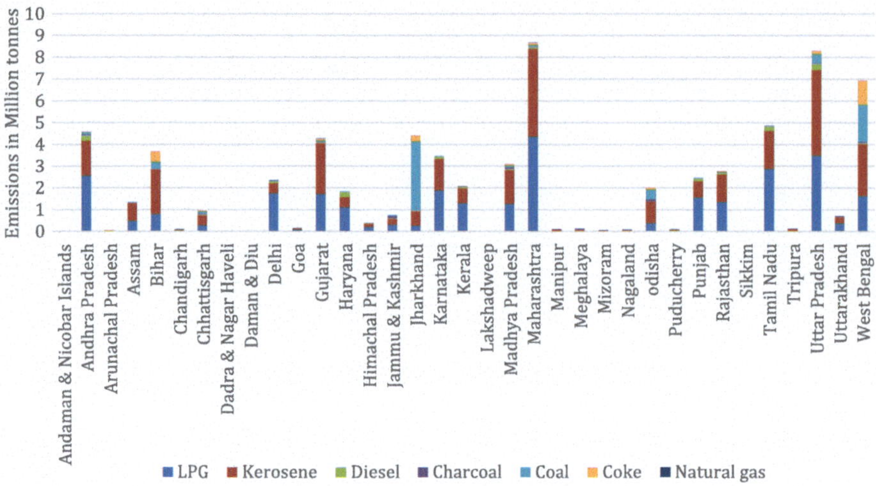

Fig. 7.3 Average primary energy HCE (2005–2010) by fuel type

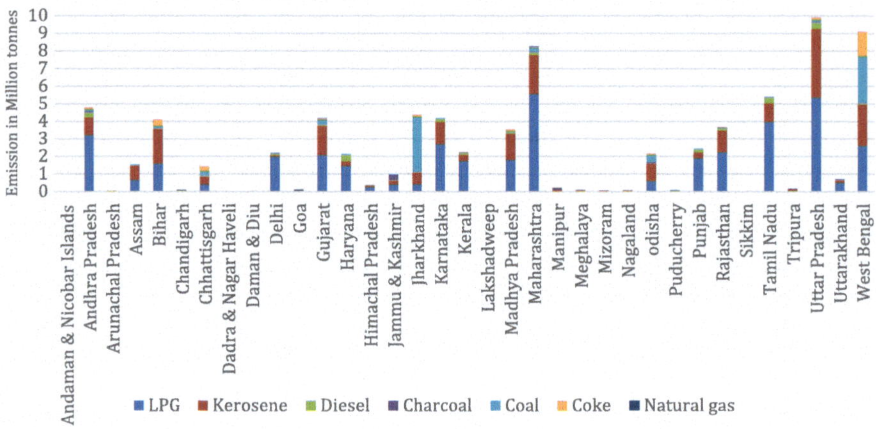

Fig. 7.4 Average primary energy HCE (2011–2015) by fuel type

LPG was promoted by the Government of India as a clean fuel. Comparing Fig. 7.3 with Fig. 7.4, it is evident that most of the States have witnessed an increase in average primary energy HCE. States like Bihar, Maharashtra, Odisha, Puducherry, Rajasthan, Tamil Nadu, Uttar Pradesh, Uttarakhand, and West Bengal are the highest emitters during the entire period under study (2005–2015). Considering the sources of fuel (LPG, diesel, charcoal, coal, coke, and natural gas), the emissions from the use of diesel, charcoal, and coke do not contribute significantly as compared to other sources of emissions.

The other significant insights from the analysis are as follows:

1. Jharkhand and West Bengal are the primary consumers of coal and have the maximum HCE due to the excessive use of the same. This may be because of the presence of coal mines in those States.

   Trends observed in average HCE due to coal during 2011–2015 over 2005–2010:
   Increasing—Gujarat (475%), Assam (449%)
   Decreasing—Tripura (−98%)

2. The average HCE due to diesel has decreased in Delhi and Goa, whereas there is a rise in the rest of the States. This can be attributed to increased consumption of LPG and natural gas for Delhi and Goa.

   Trends observed in average HCE due to diesel during 2011–2015 over 2005–2010:
   Increasing—Nagaland (45%), Arunachal Pradesh (44%)
   Decreasing—Goa (−10%)

3. There is a remarkable decline in the use of kerosene in all the States and hence a reduction in average HCE due to it. This may be due to the increased consumption of LPG as a primary fuel for cooking.

   Trends observed in average HCE due to kerosene during 2011–2015 over 2005–2010:
   Increasing—Lakshadweep (13%)
   Decreasing—Delhi (−498%), Chandigarh (−200%)

4. The highest average HCE due to charcoal is observed for the State of Jammu and Kashmir, followed by Manipur, which may be due to the climatic conditions.

   Trends observed in average HCE due to charcoal during 2011–2015 over 2005–2010:
   Increasing—Manipur (74%), Delhi (71%)
   Decreasing—Gujarat (−3843%), Chhattisgarh (−1247%)

5. West Bengal has the highest average emissions from coke, whereas there is an exponentially increasing trend in emissions observed for the State of Chhattisgarh. Maharashtra has shown a significant decline in average emissions from coke.

   Trends observed in average HCE due to coke during 2011–2015 over 2005–2010:
   Increasing—Chhattisgarh (85%), Andhra Pradesh (75%)
   Decreasing—Himachal Pradesh (−273%), Maharashtra (−225%)

6. The natural gas emissions are seen only in four States—Gujarat, Maharashtra, Delhi, and Tripura. This clearly implies the limited availability and use of natural gas as a fuel resource in the rest of the States.

Trends observed in average HCE due to natural gas during 2011–2015 over 2005–2010: Increasing—Delhi (60%), Gujarat (36%), Maharashtra (26%), Tripura (23%)

7. Average emissions due to LPG are the highest compared to other sources for the entire period of study (2005–2015).

   Trends observed in average HCE due to LPG during 2011–2015 over 2005–2010:
   Constant use—Lakshadweep (~0%)
   Increasing—Bihar (49%), Rajasthan (39%)

7.3 Cluster Analysis

Cluster analysis is a type of unsupervised learning algorithm whose objective is to form a cluster of observations that are homogeneous within and heterogeneous between. There are a number of methods used for creating such clusters. Some of the popular methods are hierarchical clustering and non-hierarchical clustering. Hierarchical clustering consists of agglomerative and divisive methods, whereas non-hierarchical clustering comprises k-means, probabilistic clustering, and the methods of self-organizing.

7.3.1 Clustering Approach

One of the disadvantages of hierarchical clustering is the difficulty in determining the number of clusters. Whereas pre-determining the number of clusters (k) is one of the challenges that is associated with k-means clustering. However, a combination of both hierarchical clustering and k-means clustering is used in most of the research work [2].

Therefore, the two-step approach is proposed to address the defined problem statement.

Step 1: Using agglomerative hierarchical clustering to get an idea about the initial clusters.

Step 2: Using k-means to actually find and interpret the clusters.

A. **Agglomerative Hierarchical Clustering**

In agglomerative method, each observation is treated as a separate cluster in the first step and then, based on a distance metric, e.g., Euclidean, grouped into clusters, eventually resulting in a single cluster. The similarity between the clusters is determined by the linkage methods. Average linkage is one of the linkage methods, which considers each pair of objects and computes its average to choose similar clusters. The data used for clustering, as shown in Table 7.1, consists of primary energy

Greenhouse Gas (GHG) emissions of 35 States and Union Territories of India for the year 2015. This dataset includes the information about primary energy household emissions i.e., CO_2, CH_4 and N_2O. CO_2 emissions are from LPG, Kerosene and Diesel, whereas CH_4 and N_2O from LPG, Kerosene, Diesel and Fuelwood. As the primary energy household emissions from kerosene are sub-zero for Delhi in 2015, the dataset is processed a bit to replace them with zero for applying clustering algorithms.

The standardized dataset is used to create clusters using the agglomerative hierarchical technique with the help of SPSS software. The output is in the form of a dendrogram, which gives an idea about choosing the initial number of clusters. The dendrogram is presented in Fig. 7.5, which clearly highlights the possibility of forming 03 clusters. In order to confirm the number of clusters, the k-means method is carried out, which is described in the next section.

B. **K-means clustering**

After getting a basic idea about the initial number of clusters using agglomerative hierarchical clustering, k-means clustering is performed on the standardized data. The k-means algorithm strongly depends on the value of k. k-means is a type of iterative algorithm which partitions the data into pre-defined k non-overlapping clusters (subgroups). The within-cluster data points are as similar as possible, and the between-cluster data points are as dissimilar as possible. These points are allocated to clusters so that the sum of the squared distance between the points and the cluster's centroid is as small as possible. There is not much difference in the centroids for the two clusters when $k = 4$, and therefore $k = 3$ is chosen as it depicts sufficient distance between the cluster centroids.

Now the dataset is divided into three clusters, but it is essential to find out the importance of the different fuel types. This can be achieved through Analysis of Variance (ANOVA), which is presented in Table 7.2. Larger the F-value, the more the variable has a role in the creation of a cluster as it is a ratio of "between-cluster variance" and "within-cluster variance". It can be observed that kerosene ($F = 49.051$) and LPG ($F = 48.880$) are the most important fuel types which are contributing to the cluster formation by minimizing within-cluster variance and maximizing between-cluster variance, followed by diesel ($F = 24.683$) and fuelwood ($F = 17.683$).

Table 7.3 describes the clusters of the States and their distances from the centroid of that cluster. It can be seen that Cluster-1 has only one State, i.e., Uttar Pradesh; Cluster-2 contains 23 States, which makes it the biggest cluster; and Cluster-3 contains 11 States.

Table 7.4 describes the cluster centers for each variable, which helps to identify the characteristics of each cluster, and is computed by taking the average of variables within each cluster.

It is clear from Fig. 7.6 and Table 7.4 that Cluster-1 has the highest cluster means (i.e., the cluster with the most emissions), followed by Cluster-3 and Cluster-2.

Table 7.1 Emissions by fuel and gas type for the year 2015

States	Fuel and gas Type										
	LPG_CO_2 (t)	Kerosene_CO_2 (t)	Diesel_CO_2 (t)	LPG_CH_4 (t)	Kerosene_CH_4 (t)	Diesel_CH_4 (t)	Fuelwood_CH_4 (t)	LPG_N_2O (t)	Kerosene_N_2O (t)	Diesel_N_2O (t)	Fuelwood_N_2O (t)
Andaman & Nicobar Islands	23,614.340	14,022.414	3222.745	1.871	1.950	0.435	228.912	0.037	0.117	0.026	3.052
Andhra Pradesh	2,527,824.0	601,059.633	175,696.07	200.303	83.597	23.711	55,216.076	4.006	5.016	1.423	736.214
Arunachal Pradesh	46,953.744	25,927.583	3037.160	3.721	3.606	0.410	2095.024	0.074	0.216	0.025	27.934
Assam	830,428.37	790,944.868	37,632.004	65.803	110.006	5.079	35,057.363	1.316	6.600	0.305	467.432
Bihar	2,157,525.2	1,953,467.575	30,617.707	170.961	271.692	4.132	41,644.468	3.419	16.302	0.248	555.260
Chandigarh	105,172.08	3013.701	6347.739	8.334	0.419	0.857	13.010	0.167	0.025	0.051	0.173
Chhattisgarh	530,284.57	395,123.430	84,643.371	42.019	54.955	11.423	77,411.980	0.840	3.297	0.685	1032.160
Dadra and Nagar Haveli	20,243.589	4625.086	2881.625	1.604	0.643	0.389	1053.442	0.032	0.039	0.023	14.046
Daman and Diu	17,331.717	2050.520	1666.118	1.373	0.285	0.225	174.105	0.027	0.017	0.013	2.321
Delhi	2,100,285.1	0.000	132,165.244	166.425	0.000	17.836	24.666	3.329	0.000	1.070	0.329
Goa	128,132.85	12,476.414	6113.814	10.153	1.735	0.825	167.056	0.203	0.104	0.050	2.227
Gujarat	2,304,242.9	1,618,452.339	74,003.114	182.587	225.098	9.987	43,430.952	3.652	13.506	0.599	579.079
Haryana	1,715,571.189	184,372.179	383,628.608	135.941	25.643	51.772	8399.191	2.719	1.539	3.106	111.989
Himachal Pradesh	331,993.565	59,332.840	33,463.318	26.307	8.252	4.516	10,824.223	0.526	0.495	0.271	144.323
Jammu and Kashmir	480,025.557	200,092.818	36,757.494	38.037	27.829	4.961	22,365.678	0.761	1.670	0.298	298.209
Jharkhand	566,747.966	642,522.279	41,606.112	44.909	89.363	5.615	22,983.708	0.898	5.362	0.337	306.449
Karnataka	3,216,717.5	1,199,073.963	232,140.06	254.890	166.770	31.328	40,374.807	5.098	10.006	1.880	538.331

(continued)

Table 7.1 (continued)

States	Fuel and gas Type										
	LPG_CO_2 (t)	Kerosene_CO_2 (t)	Diesel_CO_2 (t)	LPG_CH_4 (t)	Kerosene_CH_4 (t)	Diesel_CH_4 (t)	Fuelwood_CH_4 (t)	LPG_N_2O (t)	Kerosene_N_2O (t)	Diesel_N_2O (t)	Fuelwood_N_2O (t)
Kerala	1,954,136.4	290,052.523	130,606.68	154.844	40.341	17.626	48,884.262	3.097	2.420	1.058	651.790
Lakshadweep	753.112	2426.813	712.130	0.060	0.338	0.096	147.948	0.001	0.020	0.006	1.973
Madhya Pradesh	2,099,144.5	1,426,111.672	107,866.305	166.335	198.347	14.557	41,225.795	3.327	11.901	0.873	549.677
Maharashtra	6,082,310.7	1,554,726.415	161,134.239	481.958	216.235	21.746	48,195.753	9.639	12.974	1.305	642.610
Manipur	68,343.058	58,083.041	2268.822	5.415	8.078	0.306	2616.868	0.108	0.485	0.018	34.892
Meghalaya	42,166.144	62,538.281	7753.979	3.341	8.698	1.046	3682.991	0.067	0.522	0.063	49.107
Mizoram	63,268.202	16,740.427	1472.888	5.013	2.328	0.199	1001.379	0.100	0.140	0.012	13.352
Nagaland	52,325.205	41,261.876	1535.340	4.146	5.739	0.207	2450.741	0.083	0.344	0.012	32.677
Odisha	857,933.311	950,500.084	33,644.394	67.982	132.198	4.540	60,263.074	1.360	7.932	0.272	803.508
Puducherry	95,270.345	10,323.844	16,494.181	7.549	1.436	2.226	109.583	0.151	0.086	0.134	1.461
Punjab	2,166,438	207,369.972	188,412.272	171.667	28.841	25.427	3879.400	3.433	1.730	1.526	51.725
Rajasthan	2,738,115.9	1,189,684.075	155,749.933	216.966	165.464	21.019	98,410.710	4.339	9.928	1.261	1312.143
Sikkim	32,652.942	14,387.975	1476.382	2.587	2.001	0.199	162.527	0.052	0.120	0.012	2.167
Tamil Nadu	4,574,697.1	841,742.960	351,082.294	362.496	117.071	47.380	48,876.172	7.250	7.024	2.843	651.682
Tripura	101,270.18	94,586.256	2359.656	8.025	13.155	0.318	9931.975	0.160	0.789	0.019	132.426
Uttar Pradesh	6,632,206.1	3,832,676.132	414,529.909	525.531	533.056	55.942	97,414.256	10.511	31.983	3.357	1298.857
Uttarakhand	605,056.09	86,242.838	43,178.918	47.944	11.995	5.827	21,346.028	0.959	0.720	0.350	284.614
West Bengal	3,295,888	2,320,410.94	97,382.70	261.164	322.728	13.142	58,991.661	5.223	19.364	0.789	786.555

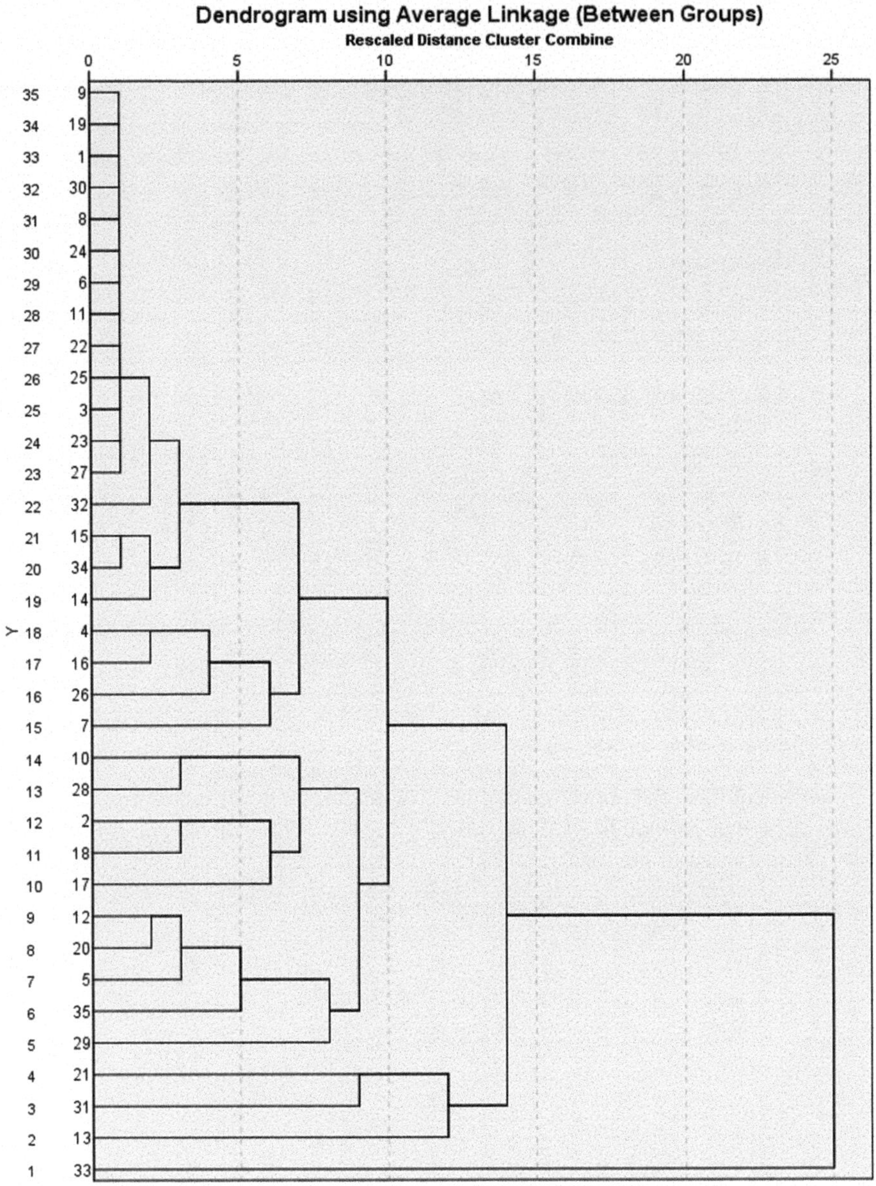

Fig. 7.5 Dendrogram

Table 7.2 Analysis of variance

ANOVA						
	Cluster		Error		F	Sig
	Mean square	df	Mean square	df		
LPG_CO_2	12.808	2	0.262	32	48.880	0.000
Kerosene_CO_2	12.819	2	0.261	32	49.051	0.000
Diesel_CO_2	10.314	2	0.418	32	24.683	0.000
LPG_CH_4	12.808	2	0.262	32	48.880	0.000
Kerosene_CH_4	12.819	2	0.261	32	49.051	0.000
Diesel_CH_4	10.314	2	0.418	32	24.683	0.000
Fuelwood_CH_4	8.925	2	0.505	32	17.683	0.000
LPG_N_2O	12.808	2	0.262	32	48.880	0.000
Kerosne_N_2O	12.819	2	0.261	32	49.051	0.000
Diesel_N_2O	10.314	2	0.418	32	24.683	0.000
Fuelwood_N_2O	8.925	2	0.505	32	17.683	0.000

7.3.2 *Results and Discussion on Profiling of Clusters*

The most significant fuel types contributing to the creation of the clusters are kerosene and LPG. Cluster-1 can be named as the "Excessive Emitters" with significantly higher emissions in comparison to the other clusters. "Uttar Pradesh" is the only State in this cluster due to the highest cluster means. Cluster-2, consisting of 23 States, as seen in Table 7.3, has the lowest cluster means (Table 7.4), which can be called as "Low Emitters". Cluster-3 has 11 States, whose cluster means fall between the "Excessive Emitters" and "Low Emitters" groups, which can be named as "High Emitters". These three clusters have been created purely on the basis of primary energy household emissions. However, it would be pragmatic to carry out the profiling of these clusters to have better insights.

A. **Evaluation of the Clusters by Population**

The projected population data for the profiling is taken from the Census of India [3]. It can be observed from Fig. 7.7 that the Lower Emissions Group has a lower population, and High Emissions States have a higher population. The Excessive Emissions Group consists of only Uttar Pradesh which has the highest population among the clusters and Indian States. It implies that population is one of the variables that explains the formation of such clusters. However, it is observed that Kerala and Haryana are members of the high-emission clusters, but they have a relatively lower population in comparison to other States of that cluster.

The insights derived from the above graph shown in Fig. 7.7 are revalidated in Fig. 7.8, which describes the members of the clusters in the form of bubbles on the map of India. While the bubble size depicts the population of the State, the color indicates the cluster to which they belong. The clusters, which are represented in the

Table 7.3 Clusters and distances of the States from the centroids

Cluster-1	Distance from Cluster-1	Cluster-2	Distance from Cluster-2	Cluster-3	Distance from Cluster-3
Uttar Pradesh	0	Andaman and Nicobar	0.85482	Andhra Pradesh	1.326127
		Arunachal Pradesh	0.778529	Bihar	2.801994
		Assam	1.750447	Gujarat	1.875277
		Chandigarh	0.815299	Haryana	4.47546
		Chhattisgarh	3.31893	Karnataka	1.02376
		D & N Haveli	0.839446	Kerala	2.193559
		Daman & Diu	0.879575	Madhya P	1.442464
		Delhi	2.402249	Maharashtra	3.190349
		Goa	0.796064	Rajasthan	2.448808
		Himachal Pradesh	0.231503	Tamil Nadu	3.250747
		Jammu and Kashmir	0.522035	West Bengal	2.616776
		Jharkhand	1.134787		
		Lakshadweep	0.894724		
		Manipur	0.739031		
		Meghalaya	0.67174		
		Mizoram	0.82467		
		Nagaland	0.767789		
		Odisha	2.869728		
		Puducherry	0.753315		
		Punjab	3.027984		
		Sikkim	0.866006		
		Tripura	0.546222		
		Uttarakhand	0.553841		

form of bubbles, highlight the need for policy initiatives and restrictions to curb the primary energy household emissions from High Emission and Excessive Emission States.

B. **State-wise Per-Capita Emissions**

It may be noted that higher emissions by the States may be due to higher population, but what about per-capita emissions? Therefore, there is a need to look at and compare the per-capita emissions of these States. Figure 7.9 represents the cluster-wise and State-wise per-capita emissions and clearly shows that members of the low-emission clusters are spread across when per-capita emissions are considered. Uttar Pradesh is in the excessive emission cluster as per primary energy household emissions, but it

Table 7.4 Final cluster centers

Final cluster centers			
Fuel type and gas type	Clusters		
	Cluster-1	Cluster-2	Cluster-3
LPG_CO_2	3.02	−0.57	0.91
Kerosene_CO_2	3.78	−0.50	0.71
Diesel_CO_2	2.91	−0.49	0.77
LPG_CH_4	3.02	−0.57	0.91
Kerosene_CH_4	3.78	−0.50	0.71
Diesel_CH_4	2.91	−0.49	0.77
Fuelwood_CH_4	2.45	−0.48	0.77
LPG_N_2O	3.02	−0.57	0.91
Kerosne_N_2O	3.78	−0.50	0.71
Diesel_N_2O	2.91	−0.49	0.77
Fuelwood_N_2O	2.45	−0.48	0.77

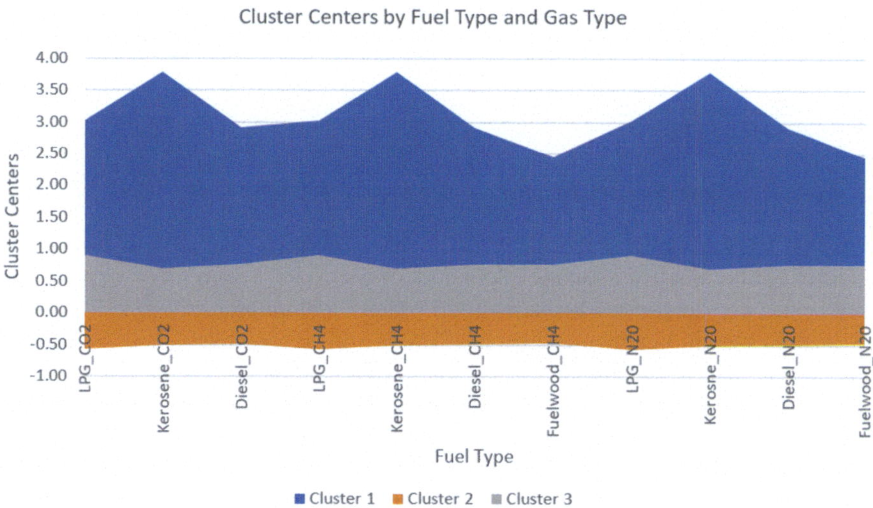

Fig. 7.6 Cluster centers by fuel type and gas type

has relatively low per-capita emissions compared to other States. Although Delhi and Punjab belong to the low-emission clusters, their per-capita emissions are the highest. In contrast, Jharkhand and Nagaland are members of the high-emission clusters, but their per-capita emissions are the lowest. Similarly, Andhra Pradesh and Bihar are part of the high-emission clusters, but their per-capita emissions are the lowest.

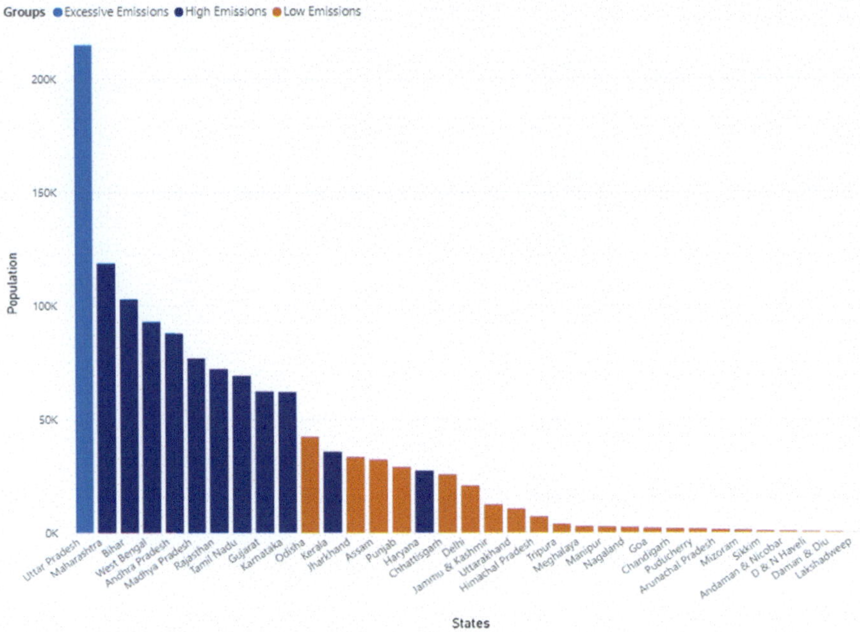

Fig. 7.7 Population by States and Groups

C. Area-wise (Per-Square Kilometer) Emissions for States

It may be worthwhile to probe the primary energy household emissions further by considering the area (per-square kilometer) of the State. Figure 7.10 depicts the cluster-wise and area-wise emissions for 35 States and union territories. It is clear that most of the union territories, except "Dadra and Nagar Haveli" and "Andaman and Nicobar Islands", are higher on per-square kilometer emissions, in spite of being a member of the low-emission clusters. Although Uttar Pradesh is a part of the excessive emission clusters, but area-wise primary energy household emissions are on the lower side.

The above insights can be incorporated while formulating policy towards curbing primary energy household emissions. Population control measures should be considered to limit the annual rise in primary energy household emissions. Another measure that can be taken is to promote the use of electric or magnetic induction and non-renewable energy in households. This simple yet effective adaptation in our daily living would help to reduce primary energy household emissions and move towards achieving the goal of zero carbon emissions.

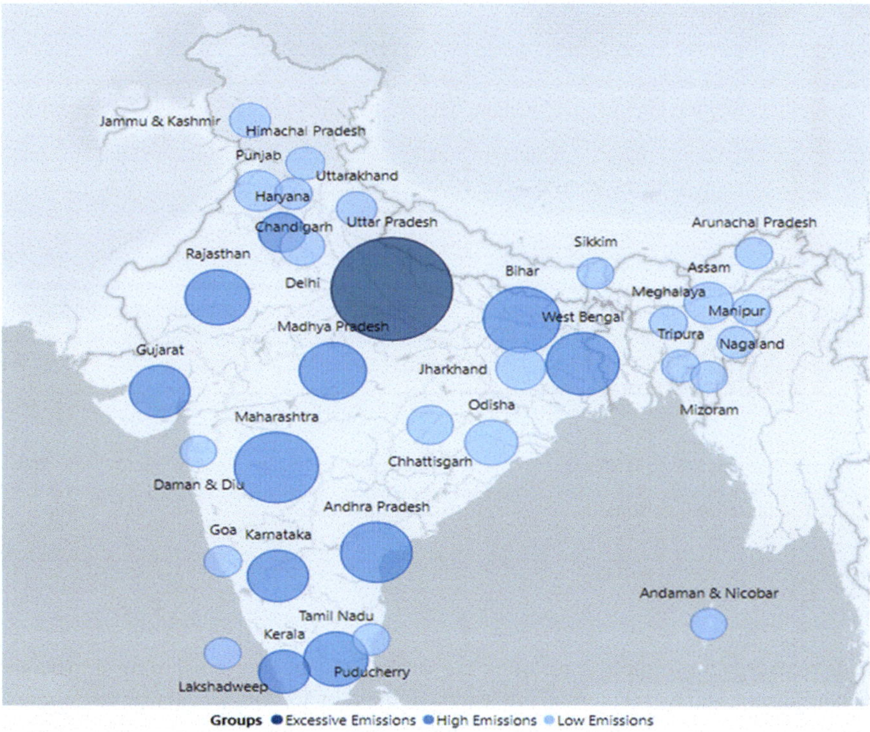

Fig. 7.8 State-wise population and emission clusters

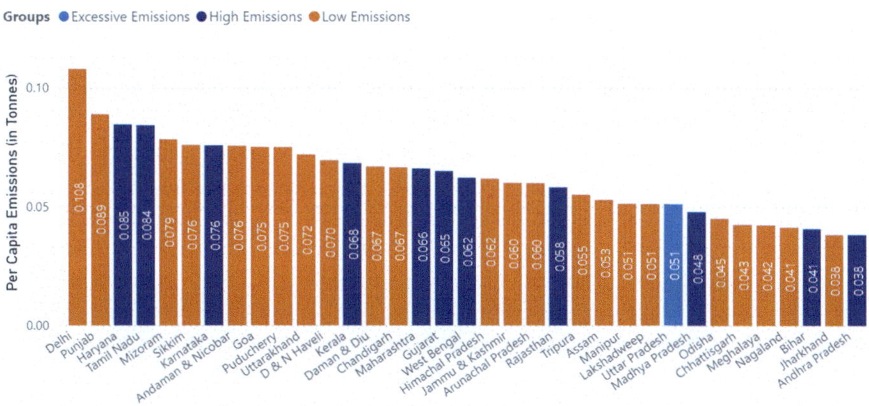

Fig. 7.9 State-wise per-capita emissions

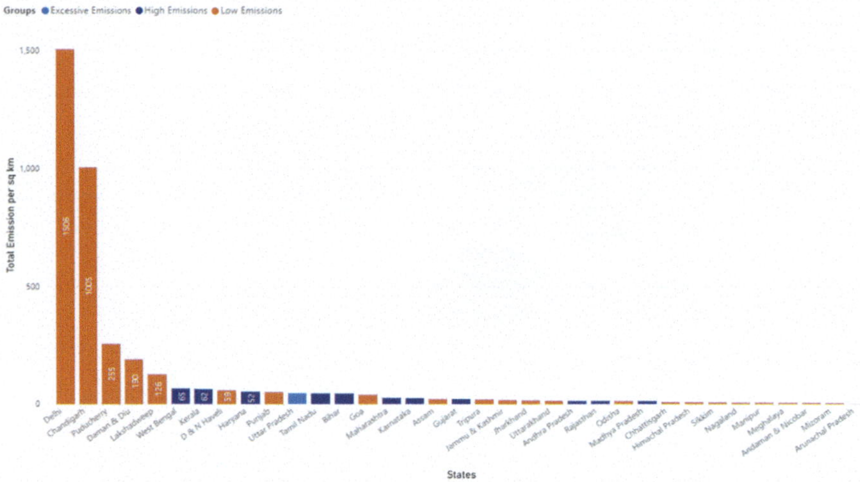

Fig. 7.10 Area-wise (per-square kilometer) emissions for States

7.4 Conclusions

Primary energy household emissions are one of the most contributing emissions from households. The residential emissions defined by the IPCC and primary energy household emissions described in various literature are similar in meaning. During 2005–2015, most of the Indian States saw an increase in primary energy HCE from LPG and a consistent decrease in primary energy HCE from kerosene. The contribution of primary energy household N_2O emissions is quite negligible for all of the States as compared to the HCE and household CH_4 emissions. As the objective of the research is to find out the groups of Indian States according to their primary energy HCE from LPG, Kerosene, Diesel, and primary energy N_2O and CH_4 from LPG, Kerosene, Diesel, and Fuelwood, the clustering methods have been applied. This created three clusters of Indian States such that they are homogeneous within and different between. "Excessive Emitters", "High Emitters", and "Low Emitters" are the three clusters formed, consisting of 1, 11, and 23 States, respectively. Kerosene and LPG are the two fuel types that have been crucial in the creation of clusters.

In order to view these clusters from different perspectives, the profiling has been carried out by considering population, per-capita emissions, and per-square kilometer emissions. It can be seen that the population of the State is one of the reasons for the level of emissions and their presence in a particular cluster. When per-capita emissions and per-square kilometer emissions are considered, the States in the low-emission cluster are found to be spread across. Uttar Pradesh is part of the excessive emission cluster, but it has relatively low per-capita emissions and per-square kilometer emissions. Most of the union territories are members of the low-emission cluster, but higher on per-square kilometer emissions. Delhi belongs to the low-emission cluster, but it is the highest both in per-capita and per-square kilometer emissions.

In light of the above insights, it is prudent to promote the use of electricity as a substitute for fossil fuels for household purposes and drive the net-zero target. Electric induction, magnetic induction stoves, etc. can be used to replace traditional cooking fuels. Many experts have advocated for a significant move towards the usage of electricity, which can be generated by non-renewable energy resources like hydropower, solar, wind, etc., as all of them are carbon-free or low-carbon sources (EcoWatch) [4]. As this study focuses on primary energy household emissions, the future scope for India in this field would be to develop a framework for calculating and finding out the important factors affecting the secondary energy HCE and household consumption HCE, also known as indirect carbon emissions in some literature.

References

1. GHG Platform India (2016) http://www.ghgplatform-india.org/energy-sector
2. Kijewska A, Bluszcz A (2016) Research of varying levels of greenhouse gas emissions in European countries using the k-means method. Atmos Pollut Res 7(5):935–944. https://doi.org/10.1016/j.apr.2016.05.010
3. Government of India, Census of India (2011) https://censusindia.gov.in/Census_Data_2001/Projected_Population/Projected_Population.pdf
4. Magnetic induction cooking can cut your kitchen's carbon footprint. EcoWatch (2020) https://www.ecowatch.com/magnetic-induction-cooking-carbon-emissions-2649690738.html

Chapter 8
Application of Machine Learning in Climate Change for Transport Sector—Literature Review and Data Preparation

8.1 Introduction

In the twenty-first century, climate change has been one of the leading problems that the whole world is trying to tackle [1]. Climate change has been defined as an imbalance in the earth's climate at a global level or even regional level over some time. It involves a change in the normal state of the atmosphere over a period that ranges from a few decennia to millions of years [1–3]. Climate change has proved to be a complex problem to deal with as the rate of climate change is much higher than expected, primarily due to industrialization and advancements across the world [2]. Although various noteworthy contributions have been made in this field by scientists and environmentalists like William Nordhaus (Nobel Prize, 2018), Michael E. Mann and Warren M. Washington (2019—Tyler Prize for Environmental Achievement), and Albert Arnold Gore Jr. (Nobel Peace Prize—2007), it requires focused action on the ground [1].

Climate change is caused primarily due to natural methods and human intervention. The natural processes include an increase in Greenhouse Gas (GHG) emissions because of different reasons like volcanic eruptions, forest fires, etc. [1, 4]. The other major natural factor is the depletion of the ozone layer, which has allowed the sun's harmful radiation to enter the earth's atmosphere [5]. These radiations trap the GHG, which increases the earth's temperature. With increased global surface temperatures, there is a high possibility of droughts and storms. Climate change has harmful impacts on the environment as well as humans. Some of the adverse effects of climate change are mentioned below:

a. NASA studied and analyzed that the temperature of the earth went up by 01% in the last century, and it is said to rise further in the next few decades. This will create disastrous effects in terms of increased temperature, more rainfall, health risks, deteriorated water quality, etc. [6].
b. According to NASA, the sea level has increased nearly by 8 inches in the last few decades. If the same trend continues, countries like Bangladesh, Sri Lanka,

N. Sharma and P. K. De, *Towards Net-Zero Targets*, Advances in Sustainability Science and Technology, https://doi.org/10.1007/978-981-19-5244-9_8

India, etc. will have most of the parts submerged underwater. This will affect the countries' environment and economy [6].

c. With the rising temperature of the earth, various changes occur in the environment, such as heatwaves and droughts, which have extensively impacted the growth of crops. This, in turn, is affecting agriculture and the economies in various countries which are heavily dependent on agriculture [6].

Human beings in the race of growth and development are also heavily contributing to the earth's declining condition. Different sectors, like energy, industrial processes, agriculture, transport, etc., are responsible for producing high volumes of GHGs and polluting natural resources such as air, water, and soil [1]. Figure 8.1 shows the contribution of different sectors towards global GHG emissions. As per the statistics shown in Fig. 8.1, the energy used in industry, buildings, and transport are the three major contributors to GHG emissions. The energy used in the industry sector accounts for 24.2% of the total GHG emissions. Whereas the energy used in buildings accounts for 17.5%, and the transport sector accounts for about 16.2% of the total GHG emissions. The unallocated fuel combustion sector accounts for 7.8%, followed by fugitive emissions from the energy sector, and livestock and manure sector, which emits 5.8% each of the total GHG emissions.

The transport sector dominates GHG emissions compared to other sectors like agriculture and the manufacturing industry. The transport sector is responsible for the

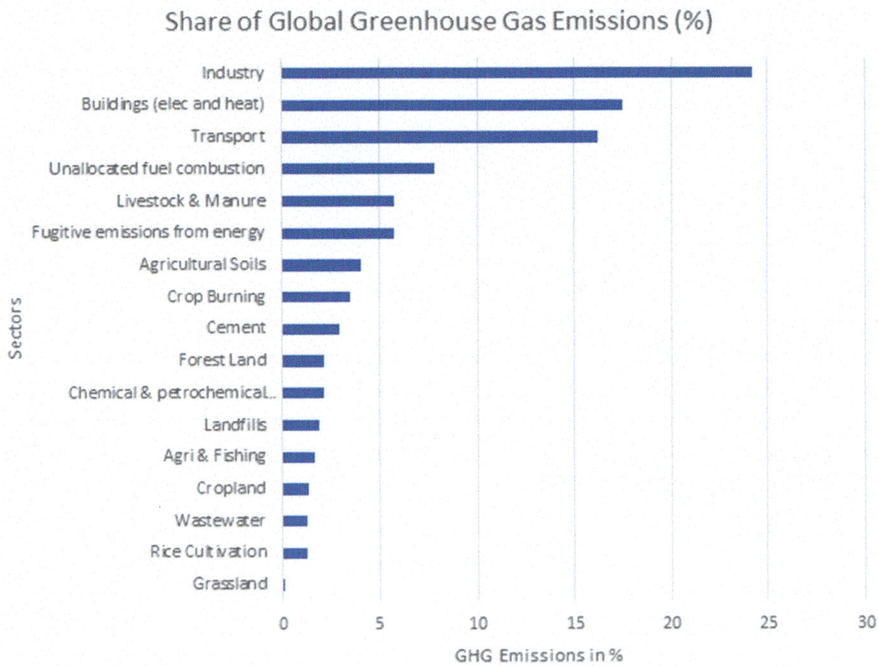

Fig. 8.1 Sector-wise distribution of greenhouse gas emission

majority of air pollutant precursor emissions [7–9]. This sector not only contributes to the CO_2 emission but also contributes towards emissions of gases like 34% of particulate matter of 2.5 microns (PM 2.5), 50% of black carbon (BC), 55% of nitrogen oxides (NO_x), 85% of carbon monoxide (CO), and 41% of non-methane volatile organic compounds (NMVOC). A significant reason for these high percentages of GHGs emitted by the transport sector is the extensive use of fossil fuels, which are used to power combustion engines [7]. But this trend is changing slowly with the increasing popularity of electric vehicles. To optimize the transport cycle, many new methods are being experimented with. Some of these methods are the implementation of various policies, the use of machine learning for optimization, social awareness camps, etc. [7, 8]. Chapters 8 and 9 aim to make use of machine learning techniques to mitigate climate change by analyzing the dataset and providing better insights. The chapters also make an attempt to predict the CO_2 emissions based on different attributes of a vehicle using machine learning.

In this chapter, an attempt is made to understand the research that has been carried out to mitigate climate change in the transport sector, identify the relevant data resource, and select the machine learning algorithms that can be applied on it. The rest of the chapter is organized as follows: Sect. 8.2 talks about the literature reviews both from systematic and contemporary perspectives, Section 8.3 focuses on the materials and methods in terms of data source and algorithms, and Sect. 8.4 concludes the chapter.

8.2 Literature Review

In this section, the endeavor is to review the literature related to climate change in the context of the transport sector. The attempt is to present a systematic literature review to understand the relevance of the topic in current times and a prevalent traditional review of the literature to understand the work done by other researchers. At the end, the research objective pertaining to transport sector and climate change is also articulated.

8.2.1 Systematic Literature Review

A systematic literature review is performed to evaluate the notable work done in the field of climate change, carbon emissions, and Machine Learning (ML) solutions to curb emissions from the transport sector. The data has been collected from the digital library of three publications, namely, Springer, IEEE Xplore, and Elsevier. The search criteria used are "Climate Change", "Climate Change due to Transport", and "ML solution for Climate Change due to Transport". Figure 8.2a–c clearly depict that there is an increasing trend in the number of papers published on climate change in the last decade, indicating the global concern related to climate change. It is evident from Fig. 8.2a–c that Elsevier has the highest number of publications in all the three search criteria compared to Springer and IEEE Xplore. An apparent rise of 170% is seen in the search criteria "Climate Change" in the year 2020 over 2011 considering the contents of all the digital libraries. Likewise, an increase of 173% and 501% is observed in the search criteria "Climate Change due to Transport" and "ML solution for Climate Change due to Transport", respectively, during the same period. It is also found that the maximum number of publications related to "ML solution for Climate Change due to Transport" is there in Elsevier compared to the other two databases. This significant surge has been noticed in the last few years in all three databases. Therefore, it is evident that Machine Learning is a vital technology that is being used and can continue to be applied in the future to tackle climate change due to the transport sector.

8.2.2 *Contemporary Literature Review*

This sub-section presents the study of literature at different levels, from understanding the impact of the transport sector as a significant contributor to climate change to the application of data science for combating the same. A comparison of published research is presented in Table 8.1 under various sections, highlighting the objective of the research, dataset, and technique used for different countries across the globe.

A. **Understanding Climate Change**

Climate change is not only influencing the atmosphere and sea levels but also changing the earth's geology. Sarin Abraham et al. researched on a similar topic and found that climate change has had a major impact on India as the average temperature across the country is predicted to go up by 1.7 °C, considering 1970 as a baseline. Not only the temperature but also the aspects like rain are expected to show anomalies. The rain is predicted to increase but with greater variability along with the risk of droughts and floods. The main sectors that are impacting climate are transport, industries, and buildings as they emit GHG emissions in a large percentage. Carbon dioxide is the main contributor to GHG emissions, which is generated by burning high carbon-producing fuels. It is predicted that the earth's temperature would increase by 2–6 °C by 2100 [10]. Consumption of fossil fuels due to transport has been a significant reason for these changes, especially in developing nations like India.

B. **Understanding Climate Change due to Transport Sector**

Many countries such as Canada, the USA, the UK, China, and Japan are investing heavily in research to find appropriate solutions for a better and more sustainable model for transportation [11]. Geffen et al., in their research, have explained that the transport sector is going to be one of the most difficult challenges concerning CO_2 emission mitigation [11, 12]. This is because of various factors like price structure,

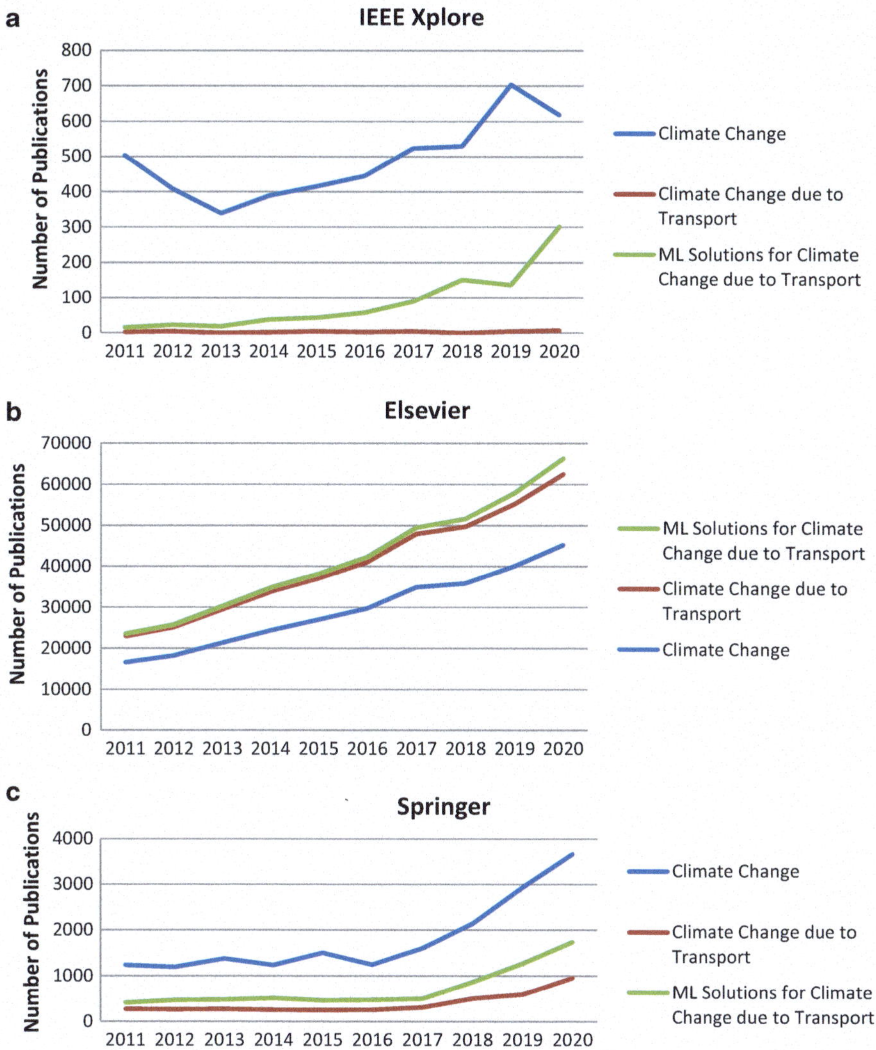

Fig. 8.2 **a** Systematic literature review of IEEE Xplore. **b** Systematic literature review of Elsevier. **c** Systematic literature review of Springer

huge infrastructure, oil as the main energy source, limited technological advancements, etc. The research has also discussed how putting a price on carbon emissions is not sufficient to reduce climate change. It was also analyzed that Business As Usual (BAU) projections for CO_2 emissions will double by the end of the century. Keeping this in mind, the paper reviewed various alternatives like hydrogen that can replace fossil fuels and several policies for carbon mitigation that can come

Table 8.1 Summary of review of literature

Area of study	Title	Year	Objective	Country	Dataset	Techniques used
Understanding climate change	Global Warming—Impacts and Future Perspective (Book)	2012	Influence of climate change on earth's geology and sectors enhancing climate change	None	None	None
Understanding climate change due to transport sector	Impact on Climate Change due to Transport Sector—Research Prospective	2012	Analyzing impacts of climate change caused by the transport sector	None	None	Statistical data analysis
	Global Climate Change and the Transportation Sector: An Update on Issues and Mitigation Options	2003	Exploring different technologies that can be used to reduce carbon emissions	None	None	None
Understanding machine learning approach to mitigate climate change	Simulation, Optimization, and Machine Learning in Sustainable Transportation Systems: Models and Applications	2021	Review various existing techniques used for designing transport systems emphasizing the sustainability factor in particular	None	None	Classification models—logistic regression, random forest, support vector machine, and neural networks

(continued)

Table 8.1 (continued)

Area of study	Title	Year	Objective	Country	Dataset	Techniques used
	An Optimized Artificial Neural Network Model using Genetic Algorithm for Prediction of Traffic Emission Concentrations	2021	Develop an OANN to enhance the existing ANN model by updating the initial weights in the network using a Genetic Algorithm (GA)	Kuala Lumpur, Malaysia	Traffic data—Ministry of Works, Malaysia, Air pollution and meteorological datasets—Department of Environment (DOE), Malaysia	Optimized Artificial Neural Network (OANN), Artificial Neural Network (ANN), genetic algorithm, random forest, and decision tree
	Machine learning for research on climate change adaptation policy integration: An exploratory UK case study	2020	Experiment with text analysis of adaptation policies using machine learning techniques and neural networks	UK	Text data https://github.com/BigDataWUR/ML4ClimateAdaptationPolicy	Text Analysis, Artificial Neural Network (ANN)
	Tackling Climate Change with Machine Learning	2019	Use machine learning to reduce greenhouse gas emissions	USA	Extreme Weather Dataset. https://paperswithcode.com/paper/extremeweather-a-large-scale-climate-dataset	Causal inference, computer vision, natural language processing, time-series analysis, unsupervised learning

(continued)

Table 8.1 (continued)

Area of study	Title	Year	Objective	Country	Dataset	Techniques used
	Machine learning model development for predicting road transport GHG emissions in Canada	2019	Prediction of GHG emission using machine learning algorithms	Canada	National Inventory Submissions 2017 https://unfccc.int/process-and-meetings/transparency-and-reporting/reporting-and-review-under-the-convention/greenhouse-gas-inventories-annex-i-parties/submissions/national-inventory-submissions-2017	Decision trees, logistic regression, multiple linear regression, ensemble techniques, Multi-Layer Perceptron (MLP)
	Prediction of Transportation Carbon Emissions using Spatio-temporal Datasets and Multilayer Perceptron Neural Network	2019	Implementing the carbon emissions calculator in real time and trying to improve its accuracy	Zhuhai, China	Zhuhai, China Real-Time Data from 5 Sources	3-Layer perceptron neural network and Internet of Things (IoT)
	A Deep Reinforcement Learning Network for Traffic Light Cycle Control	2019	Study to decide the traffic signal duration based on the collected data from different sensors using deep reinforcement learning to control the traffic light cycle	None	None	Deep reinforcement learning, Convolutional Neural Network (CNN)

(continued)

Table 8.1 (continued)

Area of study	Title	Year	Objective	Country	Dataset	Techniques used
	An Integrated Big and Fast Data Analytics Platform for Smart Urban Transportation Management	2019	Develop city administration dashboard, on top of the EUBra-BIGSEA platform that can be used by the Municipality stakeholders of Curitiba, Brazil, to tackle urban traffic data analysis and planning challenges	Curitiba, Brazil	https://ieeexplore.ieee.org/mediastore_new/IEEE/content/media/6287639/8600701/8809689/fiore.t1-2936941-large.gif	Big data, data analysis, EUBra—BIGSEA Platform (Cloud computing)
	Clustering Smart Card Data for Urban Mobility Analysis	2017	Use of smart card data generated by the AFC system to study mobility	Renne, France	https://www.star.fr/	Clustering-K means clustering, clustering based on temporal data
	Predicting Transportation Carbon Emissions with Urban Big Data	2017	Predicting carbon emissions due to transport sector using data mining techniques in real time	Zhuhai, China	Zhuhai, China Real-Time Data from 5 Sources	3-Layer perceptron neural network and Internet of Things (IoT)

(continued)

Table 8.1 (continued)

Area of study	Title	Year	Objective	Country	Dataset	Techniques used
Understanding usage of artificial intelligence to mitigate climate change	Potential of machine learning for prediction of traffic-related air pollution, Transportation Research Part D	2020	Comparing different machine learning algorithms for prediction of air quality index	Canada	https://open.canada.ca/data/en/dataset/a563e47d-6eb9-4f7f-933c-222ae49fe57f http://www.airqualityontario.com/science/data_sets.php	Artificial Neural Network (ANN) and gradient boosting Convolutional Neural Network (CNN)
	An empirical case study on Indian consumers' sentiment towards electric vehicles: A big data analytics approach	2020	Extract opinions valuable to prospective buyers, marketers, and manufacturers using deep learning techniques	India	None	Natural language processing, deep learning, support vector machine, logistic regression, and decision tree
	Topic classification of electric vehicle consumer experiences with transformer-based deep learning	2021	Deploy advances in transformer-based deep learning to discover topics of attention in a nationally representative sample of user reviews	USA	https://www.epa.gov/ghgemissions/inventory-us-greenhouse-gas-emissions-and-sinks-1990-2018	BERT, XLNet, majority classifier, LSTM, Convolutional Neural Network (CNN)

into effect. The paper concluded that climate change could be prevented by using technological advancements and de-carbonization techniques [12].

C. **Understanding Usage of Machine Learning to Mitigate Climate Change**

A review of various techniques such as artificial intelligence, machine learning, simulation, fuzzy, and optimization are used to design sustainable transportation systems by Rocio et al. [13]. In the past 10 years, it has been seen that the research for the development of sustainable transport systems has increased consistently. Simulation and optimization models remain at the forefront of development, but machine learning and fuzzy set approaches are catching up slowly. Classification models like logistic regression, random forest, support vector machines as well as neural networks are being used to find the preferences made by people for their daily commute. This further helps to make better decisions in terms of designing government schemes for promoting the use of public transport, bicycles for shorter commutes, etc. Unsupervised machine learning techniques are being used to develop autonomous electric vehicles, which are helping in reducing carbon emissions significantly [13]. Text analysis of adaptation policies using machine learning techniques and neural networks was experimented by Biesbroek et al. [14]. The research was an important part of climate change adaptation policy building and integration. Most of the time, only a few departments or organizations are considered during the formulation of the policy, such as environment, water management, agriculture, etc. The paper concluded by stating some of the pros and cons of the application of machine learning in the field of climate change and social sciences in general [14].

Rolnick et al. have done extensive research on different sectors affecting climate change and tried to solve those problems by using machine learning methods [15]. It was stated that the transport sector alone contributes to one-fourth of the CO_2 emissions. The research suggested four main approaches to tackle the GHG emissions as mentioned below:

(i) Reducing transport activity—Some of the prominent measures to minimize the transport activity are to monitor the traffic on different routes, increase the number of public transports on the roads with heavy traffic, shared mobility, freight routing, and consolidation of smaller shipments together to reduce multiple logistic movements. The authors have suggested the use of computer vision applications to monitor the volume of traffic [15].

(ii) Improvising on vehicle efficiency—Technologies like additive modeling (3D printing), drones, etc. can be of significant help in designing efficient vehicles like autonomous vehicles with hybrid engines or electric engines. The efficiency can be further enhanced using techniques like machine learning, artificial intelligence, computer vision, etc. Autonomous vehicles can be of great help in reducing CO_2 emissions as, in most of the cases, they have electric engines. It can also help in reducing traffic congestion and optimize engine performance using different modes like eco-mode, electric vehicle mode, etc. Freight transport can also benefit from autonomous vehicle systems, for example, delivery of items using drones can save miles of travel, thus contributing to the reduction of CO_2 emissions [15].

(iii) Substitution of fuel sources and electrification—The research paper covers topics like using machine learning for optimization of electric vehicles, for example, battery energy management, fault recognition in wireless charging in electric vehicles, etc. The paper also highlighted the application of machine learning in the research and development of batteries which in turn would contribute to minimizing cost and enhancing usability. It also presented some research regarding alternative fuels such as solar energy, biofuels, hydrogen, etc. and their significance in controlling climate change [15].

(iv) Modal shift—Shifting passenger and freight transport from high carbon intensity modes to low carbon intensity modes is known as modal shift. Machine learning is suggested to be used to understand passengers' travel patterns, location of pick and drop for public transports, predicting the arrival and departure of public transports, usage pattern of electric vehicles, etc. It is concluded by saying that machine learning can work as a powerful solution provider in the transport sector and can also help in de-carbonization [15].

Khan et al. performed extensive research in predicting the emission of GHG emission in Canada pertaining to the transport sector using machine learning techniques. The authors segregated the data into categorical and numerical data. The categorical data was trained using classifiers like decision tree, logistic regression, and multilayer perceptron, whereas the numerical data was trained using multiple linear regression and ensemble techniques. The best performing models were used for GHG emissions calculation [16].

D. **Understanding Usage of Artificial Intelligence to Mitigate Climate Change**

Lu et al. predicted the real-time information regarding carbon emissions of Zhuhai, China, using spatio-temporal datasets such as data from the geo positioning system of taxis, data on carbon emissions generated by the transport sector, and data related to road networks. The researchers applied a three-layer perceptron neural network that is trained on the data gathered and used the same to predict the carbon emissions caused by the transport sector. The model is evaluated using extensive experiments based on five authentic data sources from the same place. The results show that the proposed method has advantages over the well-known machine learning methods (linear regression, Gaussian naive Bayes, and logistic regression) and deep learning methods (deep belief networks and stacked denoising autoencoder) [17]. Gangavarapu has tried to implement the methodology used by Lu et al. with minor modifications for performance enhancement [18]. The authors concluded that the multilayer perceptrons provided better results than machine learning algorithms [18].

Abdullah et al. have tried to develop an Optimal Artificial Neural Network (OANN) that predicts the concentration of CO, NO, NO_2, and NO_X pollutants produced by the transport sector in the city of Kuala Lumpur. The OANN is developed by enhancing the Artificial Neural Network (ANN) by upgrading the original weights of the network with the help of a Genetic Algorithm. These results were further compared with ANN, random forest, and decision tree models. The authors concluded by stating that the OANN performed much better than the other machine learning models and had the least mean squared error of all [19]. A big data analytics platform for transport management for the urban city of Curitiba, Brazil, was developed by Fiore et al. [20]. The process involves the use of Open Data made available by the local municipality, which contains bus card information, general transit feed specification files, bus stops, and bus position data. The data is processed using applications like Apache Spark and Hadoop and analyzed using COMPSs and Ophidia frameworks. Finally, the results are made available for the users to access through a GUI dashboard. The entire project is done using cloud-based architecture. The end information is used by municipal authorities for transport optimization, managing the supply and demand of public transport in different parts of the city [20].

Mahrsi et al. have tried to use the smart card data generated by the automatic fare collection system to study the mobility of the transport system [21]. The paper presents two main approaches for clustering the data that is provided by the metropolitan area of Renne, France. This data is used to extract the mobility patterns in public transport systems. The first approach involves clustering the stations based on the trips made at a particular station, which are distributed over time. The second approach allows identifying groups of passengers with similar boarding times, which is aggregated into weekly profiles to help manage the supply and demand of public transport [21]. An attempt is made by Liang et al. to improve the efficiency of the current traffic signal control system using deep reinforcement learning [22]. The model takes the live traffic data at the intersection as input and converts it into smaller grids. The change of traffic lights in real time depending on the actual traffic is modeled as high-dimension Markov decision processes. The reward, in this case, is the waiting time between two cycles which are mapped to the different states and decided using a convolutional neural network. For evaluation of the model, the researchers have used simulation of urban mobility in a vehicular network which helps to study the efficiency of the created model in the real world [22].

Wang et al. have done a comparative study to predict air quality using traditional Land Use Regression and Machine Learning Algorithms like artificial neural networks and extreme gradient boosting algorithms [23]. The Land Use Regression method is frequently used for analyzing pollution in densely populated areas. The data is collected by air-quality sampling and a live video feed, which captures the number and types of vehicles. Every frame is first treated as an image, and the objects are boxed using multi-object detection. Once the object crosses a line in the video feed, the item is added to the database. This model was evaluated by manually counting the cars in the video [23]. Jena conducted research to understand the mindset of probable buyers of electric vehicles and help manufacturers understand the attributes that are preferred [24]. The data for this research is collected from various social media platforms, and various deep learning techniques like Recurrent Neural Network (RNN), Convolutional Neural Network (CNN), etc. are applied. The main motive of the research is to provide buyers, marketers, and industrialists with an overview of the market and make appropriate decisions accordingly [24].

Ha et al. have used transformative deep learning techniques to classify various reviews of the buyers related to the electric vehicle industry [25]. The results showed that the transformative deep learning models outperformed the typical deep learning techniques. The main objective of the research is to evaluate and bring optimization to the electric vehicle industry and also help in policy building advisory. The study concludes by stating that in future, the same technique can be applied to real-time data for better results [25]. Bacciu et al. have used various machine learning techniques to enhance the smart transportation sector [26]. The main focus of this research is bike-sharing services and optimization of these services to increase their feasibility and usage. The paper suggests the use of GPS data to access location preview and analyze the results with the help of real-time data as a future work of this project [26].

Finally, after conducting a rigorous literature review, the objective of the research is defined as—the application of machine learning techniques to predict CO_2 emissions from the transport sector to mitigate climate change.

8.3 Materials and Methods

This section discusses topics like the source and description of the dataset used in this research, along with the methods/algorithms/techniques used for data preprocessing, exploratory data analysis, and building a model.

8.3.1 Data Source Identification

To meet the objective mentioned above, there is a need for a dataset, which is well structured, covers a wide range of vehicle types, and contains no missing data. Hence, the Canadian car dataset is being used as it has all the relevant parameters of a car, which are required to predict CO_2 emissions. Besides, the amount of data available in the dataset is sufficient to create an excellent predictive model.

A. **Dataset**

The data is extracted from the official website of the Canadian government for open data [27]. The title of the data source is "Fuel Consumption Rating", which records the amount of CO_2 emissions by a vehicle depending on its various features. The compiled version of the dataset consists of data for 5 years, i.e., from 2017 to 2021. The open data provided by the Canadian government is analyzed by using Python programming language. The below code shows the first step of importing the required Python libraries:

```
[]
import numpy as np
import pandas as pd
import matplotlib.pyplot as plt
import seaborn as sns
import plotly
```

```
df=   pd.read_csv('../input/co2-emission-by-vehicles-201721/2017-
18-19-20-21 CO2 Emissions - Sheet1.csv')
```

To know the size of the dataset in terms of the number of rows and columns, the following Python code is used:

```
[In] df.shape
[Out] (5085, 15)
```

It is clear from the above code that there are 5085 rows and 15 columns in the dataset. To know the column name, "not null" values present in the mentioned columns and the data type, the following info() function is used:

```
[In] df.info()
[Out] <class 'pandas.core.frame.DataFrame'>
RangeIndex: 5085 entries, 0 to 5084

Data columns (total 15 columns):
```

#	Column	Non-Null Count	Dtype
---	------	--------------	-----
0	MODEL YEAR	5085 non-null	int64
1	MAKE	5085 non-null	Object
2	MODEL	5085 non-null	Object
3	VEHICLE CLASS	5085 non-null	Object
4	ENGINE SIZE	5085 non-null	float64
5	CYLINDERS	5085 non-null	int64
6	TRANSMISSION	5085 non-null	Object
7	FUEL TYPE	5085 non-null	Object
8	FUEL CONSUMPTION CITY (L/100 km)	5085 non-null	float64
9	FUEL CONSUMPTION HWY (L/100 km)	5085 non-null	float64
10	FUEL CONSUMPTION COMB (L/100 km)	5085 non-null	float64
11	COMB (mpg)	5085 non-null	int64
12	CO2 EMISSIONS (g/km)	5085 non-null	int64
13	CO2 RATING	5085 non-null	int64
14	SMOG RATING	5085 non-null	int64

```
dtypes: float64(4), int64(6), object(5)
memory usage: 596.0+ KB
```

It is evident from the above result that there are no "NULL" values present in the dataset. The data is mainly of three types, namely, float64, int64, and object. It is observed that most of the data values are numeric.

B. **Dataset Description and Attributes**

In the dataset, the columns describe the features, and rows contain the data related to the features. The features are "Model Year", "Make", "Model", "Vehicle Class", "Engine Size", "Cylinders", "Transmission", "Fuel Type", "Fuel Consumption City (L/100 km)", "Fuel Consumption Hwy (L/100 km)", "Fuel Consumption Comb (L/100 km)", "Comb (Mpg)", "CO_2 Emissions (G/Km)", "CO_2 Rating", "Smog Rating". A detailed description of the features and their abbreviations are mentioned in Table 8.2.

8.3.2 Algorithms and Evaluation Metrics

In this sub-section, the algorithms and the evaluation metrics used in the research related to carbon emissions in the transport sector are discussed.

Power Transformer—It is a transformation technique that helps to transform the data, similar to a normal distribution. This transformer belongs to a group of parametric, monotonic transformations, which come in handy when working with data having non-constant variance. Power transformer supports two types of statistical transformation techniques, viz., Box-Cox transform and the Yeo-Johnson transform. Box-Cox requires input data to be strictly positive, while Yeo-Johnson supports both positive and negative data.

The formula for Box-Cox transform:

$$\begin{cases} y = \frac{x^{\lambda}-1}{\lambda}, here\lambda \neq 0 \\ y = \ln x, where\lambda = 0 \end{cases}$$

Here:

y is the transformation result,
x is the variable under transformation,
λ is the transformation parameter. is the transformation parameter.

Formula for Yeo-Johnson transform:

$$\psi(y,\lambda) = \begin{cases} \dfrac{(y+1)^{\lambda}-1}{\lambda} \; if\ \lambda \neq 0, \quad y \geq 0 \\ \log(y+1) \;\; if\ \lambda = 0, \quad y \geq 0 \\ -\dfrac{[(-y+1)^{2-\lambda}-1]}{2-\lambda} \; if\ \lambda \neq 2, y < 0 \\ -\log(-y+1) \quad if\ \lambda = 2, y \ < 0 \end{cases}$$

Here:

y is the variable under transformation and,

Table 8.2 Data description

Feature name	Description	Abbreviations (if any)
Model year	The year in which the model of the vehicle is manufactured	None
Make	The company that manufactures the vehicle	None
Model	Car model	4WD/4X4 = Four-wheel drive AWD = All-wheel drive FFV = Flexible-fuel vehicle SWB = Short wheelbase LWB = Long wheelbase EWB = Extended wheelbase
Vehicle class	Class of the vehicle depending on their utility, capacity, and weight	None
Engine size (L)	Size of engine used in Liter	None
Cylinders	Number of cylinders	None
Transmission	Transmission type with number of gears	A = Automatic AM = Automated Manual AS = Automatic with Select shift AV = Continuously Variable M = Manual 3 - 10 = Number of Gears
Fuel type	Type of fuel used	X = Regular gasoline Z = Premium gasoline D = Diesel E = Ethanol (E85) N = Natural gas
Fuel consumption city (L/100 km)	Fuel consumption on city roads (L/100 km)	None
Fuel consumption HWY (L/100 km)	Fuel consumption on highways (L/100 km)	None
Fuel consumption comb (L/100 km)	The combined fuel consumption (55 percent city, 45 percent highway) is shown in L/100 km	None
Fuel consumption comb (mpg)	The combined fuel consumption in both city and highway is shown in miles per gallon (mpg)	None
CO_2 emissions (g/km)	The tailpipe emissions of carbon dioxide (in grams per kilometer) for combined city and highway driving	None
CO_2 rating	Rating of CO_2	None

(continued)

Table 8.2 (continued)

Feature name	Description	Abbreviations (if any)
SMOG rating	Rating of the SMOG	None

λ is the transformation parameter.

Quantile Transformer—This transformation technique converts the data into a normal/uniform distribution. It is a non-parametric transformation technique. It is one of the most powerful preprocessing techniques, which also helps in reducing the impact of outliers.

The formula for quantile transform:

$$y = G^{-1} f(x)$$

Here:

y is the transformation result,
f is the cumulative distribution function, and
G^{-1} is the quantile function of our output distribution.

Multiple Linear Regression—It is a machine learning algorithm used for supervised learning. It performs the task of predicting a dependent variable (target) based on the given independent variable(s).

Formula of multiple linear regression:

$$Y_i = \theta_0 + \theta_1.X_{i1} + \theta_2.X_{i2} + \ldots + \theta_p.X_{ip} + \epsilon$$

where for i = n observations.

X_i = independent variable (input),
Y_i = dependent variable (output),
θ_0 = intercept on y-axis,
θ_p = slope for the line X_i, and
ϵ = model's error term (residuals).

Random Forest Regressor—It is a supervised learning algorithm used for classification and regression. Random Forest is an ensemble (combination) of decision trees. The input data is passed through multiple decision trees. It executes by constructing a different number of decision trees at training time. The output is the mode of the classes (for classification) or mean prediction (for regression) of the individual trees.

Support Vector Regressor (SVR)—It is a supervised learning algorithm used to predict discrete values. The basic idea behind SVR is to find the best fit line. In SVR, the best fit line is the hyper-plane that has the maximum number of points.

XGBoost Regressor—It is one of the most dynamic approaches in constructing a supervised regression model. The objective function contains a loss function and a regularization term. It provides insight into the prediction rate of the model and explains the accuracy by giving the difference between the actual value and the predicted value. The primary loss function for regression problems is reg: linear, which is a learning task parameter. This parameter is used to define the optimization objective and the metric to be calculated at each step.

Lasso Regression—It is a type of linear regression that uses shrinkage. Shrinkage is where data values are shrunk towards a central point, i.e., mean. The lasso approach stimulates simple, sparse models (i.e., models with fewer parameters).

$$\text{The formula of Lasso Regression} = \underbrace{\sum_{i=1}^{n}\left(y_i - \sum_{j=0}^{p} x_{i\,j}\cdot\beta_j\right)^2}_{\text{Residual Sum of Squares}} + \underbrace{\lambda\sum_{j=0}^{p}|\beta_j|}_{\substack{\text{Sum of}\\ \text{the absolute value}\\ \text{of the magnitude}\\ \text{of coefficients}}}$$

where

λ = amount of shrinkage.

Ridge Regression—It is an example of a shrinkage method. When compared to least squares, it shrinks the parameter estimates in the hope of reducing variance, improving prediction accuracy, and aiding interpretation.

$$\text{Formula of Ridge Regression} = \underbrace{\sum_{i=1}^{n}\left(y_i - \sum_{j=0}^{p} x_{i\,j}\cdot\beta_j\right)^2}_{\text{Residual Sum of Squares}} + \underbrace{\lambda\sum_{j=0}^{p}\beta_j^2}_{\substack{\text{Sum of}\\ \text{the squared value}\\ \text{of the magnitude}\\ \text{of coefficients}}}$$

where

λ = amount of shrinkage.

Root Mean Squared Logarithmic Error (RMSLE)—It is the logarithm of the standard deviation of the prediction errors. The prediction error is a measure of distance between the regression line and data points that highlights the concentration of data around the line of best fit. Root mean square logarithmic error is commonly used in regression analysis to verify experimental results.

$$\text{Formula of Root Mean Square Logarithmic Error} = Log\left(\sqrt{\frac{\sum_{i=1}^{N}(P_i - A_i)^2}{N}}\right)$$

where

P_i = predicted value,
A_i = actual value, and
N = number of observations.

R^2 Score—R-squared is a goodness-of-fit measure for linear regression models. This score indicates the percentage of the variance in the dependent variable that the independent variables explain collectively. R-squared measures the strength of the relationship between the model and the dependent variable on a convenient scale of 0–1.

$$\text{Formula of } R^2\text{score} = 1 - \frac{\sum(y_i - \hat{y})^2}{\sum(y_i - \overline{y})^2}$$

where

$\hat{y}$ = predicted value of y and
$\overline{y}$ = mean value of y.

8.4 Conclusions

In this chapter, an attempt has been made to review various works of literature from different perspectives, like how transport and climate change are interrelated. Various research works carried out in the transport sector with the help of machine learning and deep learning algorithms are reviewed. From the literature, it is observed that countries like the USA, the UK, Canada, and France are working at the forefront to solve the climate change problem caused by the transport sector. The popular machine learning algorithms used by the authors are k-nearest neighbor (KNN), decision tree, logistic regression, etc. Besides, with the advancement in technology, the application of deep learning techniques, natural language processing (NLP), artificial neural network (ANN), long short-term memory (LSTM), and many other algorithms are used to solve the problem of climate change to achieve the net-zero goal. In the later section of this chapter, a dataset has been identified for the identified problem statement. Finally, there is a discussion on the material and methods that are used for exploratory data analysis and building a model.

References

1. Mohanty S, Mohanty B (2009) Global climate change: a cause of concern
2. IPCC (2013) AR5 climate change 2013: the physical science basis—IPCC
3. Intergovernmental Panel on Climate Change (2014)
4. Baliram L, Jadhav G (2020) Climate change
5. US Global Change Research Program. References in Global climate change and children's health—GCIS
6. ReliefWeb (2019) WMO greenhouse gas bulletin: the state of greenhouse gases in the atmosphere based on global observations through 2018 (No. 15 | 25 November 2019)—World
7. Gordon D (2010) The role of transportation in driving climate disruption energy and climate program
8. Unger N, Shindell DT, Wang JS (2009) Climate forcing by the on-road transportation and power generation sectors. Atmos Environ 43(19):3077–3085
9. Fuglestvedt J, Berntsen T, Myhre G, Rypdal K, Skeie RB (2008) Climate forcing from the transport sectors. Proc Natl Acad Sci 105(2):454–458
10. Singh BR, Singh O (2012) Study of impacts of global warming on climate change: rise in sea level and disaster frequency
11. Abraham S, Ganesh K, Kumar AS, Ducqd Y (2012) Impact on climate change due to transportation sector—research prospective. Procedia Engineering 38:3869–3879
12. Geffen C, Dooley J, Kim S (2003) Global climate change and the transportation sector: an update on issues and mitigation options
13. de la Torre R, Corlu CG, Faulin J, Onggo BS, Juan AA (2021) Simulation, optimization, and machine learning in sustainable transportation systems: models and applications. Sustainability 13(3):1551
14. Biesbroek R, Badloe S, Athanasiadis IN (2020) Machine learning for research on climate change adaptation policy integration: an exploratory UK case study. Regional Environmental Change 20(3)
15. Rolnick D, Donti Priya L, Kaack LH, Kochanski K, Lacoste A, Sankaran K, Ross AS, Milojevic-Dupont N, Jaques N, Waldman-Brown A, Luccioni A, Maharaj T, Sherwin ED, Karthik MS, Kording Konrad P, Gomes C, Ng AY, Hassabis D, Platt JC, Creutzig F (2019) Tackling climate change with machine learning
16. Khan MJUR, Awasthi A (2019) Machine learning model development for predicting road transport GHG emissions in Canada. WSB J Bus Financ 53(2):55–72
17. Lu X, Ota K, Dong M, Yu C, Jin H (2017) Predicting transportation carbon emission with urban big data. IEEE Trans Sustain Comput 2(4):333–344
18. Mahesh G, Reddy M (2018) Prediction of transportation carbon emission using Spatio-temporal datasets and multilayer perceptron neural network. Int J New Innov Eng Technol 8(2)
19. Abdullah AM, Usmani RSA, Pillai TR, Marjani M, Hashem IAT (2021) An optimized artificial neural network model using genetic algorithm for prediction of traffic emission concentrations. Int J Adv Comput Sci Appl (IJACSA)
20. Fiore S, Elia D, Pires CE, Mestre DG, Cappiello C, Vitali M, Andrade N, Braz T, Lezzi D, Moraes R, Basso T, Kozievitch NP, Fonseca KVO, Antunes N, Vieira M, Palazzo C, Blanquer

I, Meira W, Aloisio G (2019) An integrated big and fast data analytics platform for smart urban transportation management. IEEE Access 7:117652–117677
21. El Mahrsi MK, Come E, Oukhellou L, Verleysen M (2017) Clustering smart card data for urban mobility analysis. IEEE Trans Intell Transp Syst 18(3):712–728
22. Liang X, Du X, Wang G, Han Z (2019) A deep reinforcement learning network for traffic light cycle control. IEEE Trans Veh Technol 68(2):1243–1253
23. Wang A, Xu J, Tu R, Saleh M, Hatzopoulou M (2020) Potential of machine learning for prediction of traffic related air pollution. Transp Res Part D: Transp Environ 88:102599
24. Jena R (2020) An empirical case study on Indian consumers' sentiment towards electric vehicles: a big data analytics approach. Industrial Marketing Management
25. Ha S, Marchetto DJ, Dharur S, Asensio OI (2021) Topic classification of electric vehicle consumer experiences with transformer-based deep learning. Patterns 2(2):100195. https://doi.org/10.1016/j.patter.2020.100195
26. Bacciu D, Carta A, Gnesi S, Semini L (2017) An experience in using machine learning for short-term predictions in smart transportation systems. J LogAl Algebr Methods Program 87:52–66
27. https://open.canada.ca/data/en/dataset/98f1a129-f628-4ce4-b24d-6f16bf24dd64#wb-auto-6

Chapter 9
Application of Machine Learning to Predict CO_2 Emission from Transport Sector to Mitigate Climate Change

9.1 Introduction

In Chap. 8, climate change due to the transport sector was explored by reviewing various literature from different perspectives, like understanding climate change, its impact, and the transport sector as a significant contributor to the usage of data science for combating the same. The study process helped in identifying the objectives for the current research, reaching out to the most appropriate open dataset to meet the objectives. It also provided the detailed information regarding the dataset, its attributes and relevant techniques for building the analytical model.

The chapter presents the exploratory data analysis to describe the dataset and find the relationship between the two attributes. The chapter further explains the step-by-step process of applying machine learning algorithms to predict CO_2 emissions using the various attributes of a vehicle from the dataset. The rest of the chapter is organized as follows: Sect. 9.2 presents data preparation and exploratory data analysis (EDA). The section focuses on an analysis of data from various perspectives, presents the correlation between the attributes, and discusses the feature selection for building up a model. Section 9.3 deliberates on building up various models, evaluates their performance, and suggests the best model for predicting CO_2 emissions. Finally, Sect. 9.4 presents the conclusions of the chapter.

9.2 Data Preparation and Exploratory Data Analysis (EDA)

This section describes various trends and correlations between variables by carrying out exploratory data analysis. The variables taken under consideration are continuous, discrete, and categorical. The continuous variables are the variables that can take an infinite variety of values between the lowest and the highest points of measurement. Whereas discrete and categorical variables are similar in the sense that they can only

N. Sharma and P. K. De, *Towards Net-Zero Targets*, Advances in Sustainability Science and Technology, https://doi.org/10.1007/978-981-19-5244-9_9

take finite values between the lowest and the highest points of measurement. The only difference between them is that categorical variables do not have any intrinsic order.

The variables have been described and analyzed using various libraries. The libraries used for creating all the visualization are Matplotlib, Seaborn, and Plotly. These libraries are open-source plotting libraries available in Python. Matplotlib is a 2D plotting library that helps in visualizing figures. Seaborn is an upgraded version of Matplotlib that helps in visualizing the data in a better way. Plotly supports multiple programming languages like MATLAB, Python, R, and JavaScript and is great for building beautiful and interactive visualizations.

9.2.1 Data Preparation

In this section, distribution of continuous variables such as "FUEL CONSUMPTION CITY (L/100 km (Liter/100 km))", "FUEL CONSUMPTION HWY (L/100 km (Liter/100 km))", "FUEL CONSUMPTION COMB (L/100 km (Liter/100 km))", "COMB (mpg—miles per gallon)" is studied.

Note: FUEL CONSUMPTION COMB (L/100 km = The combined fuel consumption (55% city, 45% highway) is shown in L/100 km. COMB (mpg) = The combined fuel consumption in both city and highway is shown in mile per gallon (mpg).

The code snippet shown below is used to plot the boxplots for the continuous variables to identify the outliers:

```
columns = ['FUEL CONSUMPTION CITY (L/100 km)', 'FUEL CONSUMPTION HWY
(L/100 km)', 'FUEL CONSUMPTION COMB (L/100 km)', 'COMB (mpg)']
def box_plots(columns, bins = 50):
fig = plt.figure(figsize=(15, 10))
plt.suptitle("BOX PLOTS")
counter = 1
for i in columns:
plt.subplot(2,2, counter)
sns.set_theme(style="whitegrid")
sns.boxplot(x=df[i], color='lightblue')
counter += 1
box_plots(columns)
```

After executing the above code, the boxplot is generated as shown in Fig. 9.1.

Figure 9.1 shows that all the continuous variables have some outliers. An outlier is a value that deviates enormously from other values in a random sample of a population. Outliers can increase the variability in the data, which decrease its statistical power and introduce bias in a machine learning model. From Fig. 9.1, it is observed that "FUEL CONSUMPTION CITY" has outliers beyond 22 L/100 km, "FUEL CONSUMPTION HWY" has outliers above 14 L/100 km, "FUEL CONSUMPTION COMB" has outliers beyond 18 L/100 km, and "COMB" has outliers above 44 mpg.

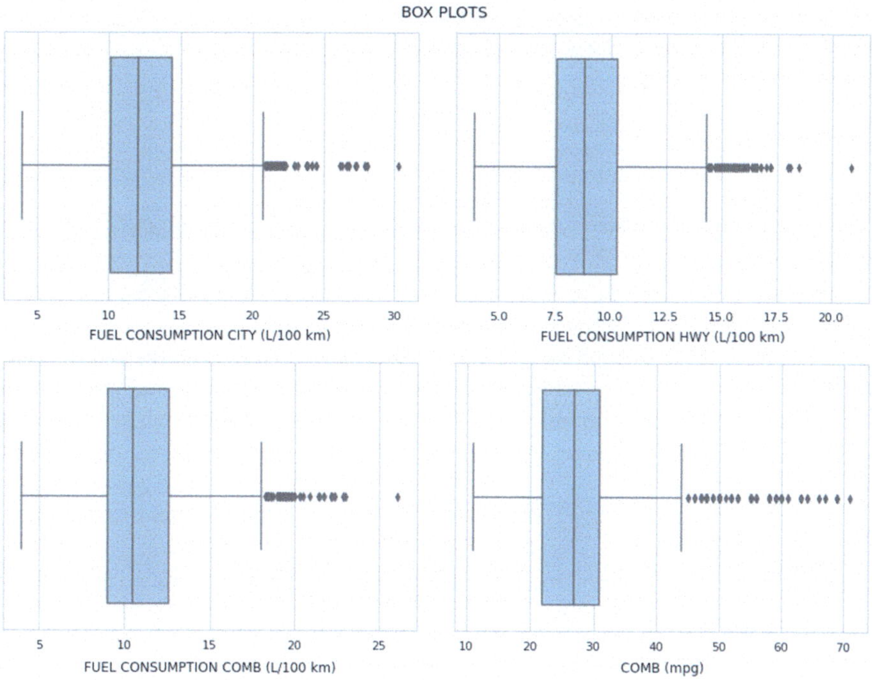

Fig. 9.1 Boxplots of continuous variables in the dataset

To remove these outliers, two data transformation techniques are used, namely, power and quantile transformers (explanation provided in Chap. 8, Sect. 8.3.1C). These transformers are used to make the data more similar to a normal distribution. The fragment of code below compares the original data distribution with the distribution after applying power and quantile transformers.

```
from sklearn.preprocessing import PowerTransformer, QuantileTrans-
former
def check_best_transformer(columns):
pt = PowerTransformer()
qt = QuantileTransformer(n_quantiles=100,output_distribution=
'normal') fig = plt.figure(figsize=(20,30))
plt.suptitle("Quantitle Vs Power Transform")
sns.set_theme(style="whitegrid")
j = 1
for i in columns:
array = np.array(df[i]).reshape(-1, 1)
y = pt.fit_transform(array)
x = qt.fit_transform(array)
plt.subplot(4,3,j)
sns.histplot(array, bins = 30, kde = True)
plt.title(f"Original Distribution for {i}")
plt.subplot(4,3,j+1)
```

```
sns.histplot(x, bins = 30, kde = True)
plt.title(f"Quantile Transform for {i}")
plt.subplot(4,3,j+2)
sns.histplot(y, bins = 30, kde = True)
plt.title(f"Power Transform for {i}")
j += 3
check_best_transformer(columns)
```

After executing the above code, the histograms are generated as shown in Fig. 9.2a, b, c and d.

Now, from Fig. 9.2a, b, c, and d, it is seen that the quantile transformer performs slightly better than the power transformer while considering the continuous variables (“FUEL CONSUMPTION CITY”, “FUEL CONSUMPTION HYW”, “FUEL CONSUMPTION COMB”, “COMB”). The quantile transformer normalizes the data by converting it to a uniform bell curve of the normal distribution, when compared to a power transformer. Hence, quantile transformer is used to remove the outliers from the dataset.

The fragment of code below is used to apply the quantile transformer to the dataset:

```
qt = QuantileTransformer(n_quantiles=100, output_distribution=
'normal') for i in columns:
array = np.array(df[i]).reshape(-1, 1)
y = qt.fit_transform(array)
df[i] = y
```

9.2.2 *Analysis of Various Features*

Frequency analysis is performed to understand the trends in each feature. Frequency analysis is a statistical technique that helps in analyzing the data by measuring its frequency. This technique is performed on categorical variables like “MAKE”, “VEHICLE CLASS”, “TRANSMISSION”, “FUEL TYPE”, and discrete variables such as “ENGINE SIZE”, “CYLINDERS”, “CO_2 RATING”, “SMOG RATING”.

A. **Top 20 Companies with Highest Number of Vehicles Manufactured**

The general observation is that the popularity of any vehicle is directly proportional to the quality of the vehicle. Better quality not only translates to better features and safety standards but also to more efficient engines that produce lower carbon emissions. Hence, it makes it necessary to study the popular car manufacturers in Canada. The following code helps in visualizing the top 20 companies with the highest number of vehicles:

```
import plotly.express as px
import plotly.graph_objects as go
from plotly.subplots import make_subplots
df_Make=df['MAKE'].value_counts().reset_index().rename(columns=
```

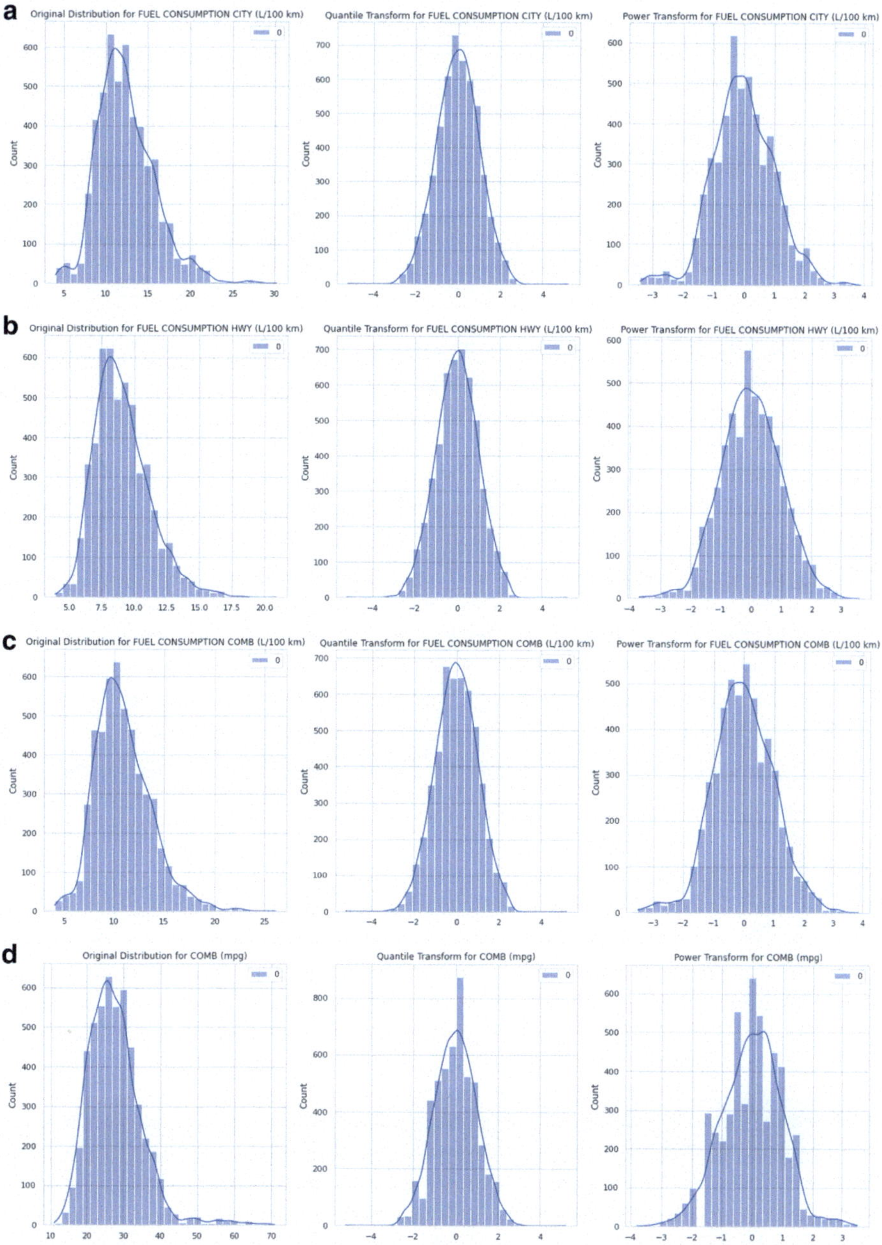

Fig. 9.2 **a** Comparison between power transformer and quantile transformer applied on "FUEL CONSUMPTION CITY". **b**. Comparison between power transformer and quantile transformer applied on "FUEL CONSUMPTION HYW". **c** Comparison between power transformer and quantile transformer applied on "FUEL CONSUMPTION COMB". **d**. Comparison between power transformer and quantile transformer applied on "COMB"

```
{'index':'MAKE','MAKE':'Count'})[0:20] df_Make
fig = go.Figure(go.Bar(
x=df_Make['MAKE'],y=df_Make['Count'],
marker={'color': df_Make['Count'],
'colorscale': 'Sunset'},
text=df_Make['Count'],
textposition = "outside",
))
fig.update_layout(title_text='Top 20 Company',xaxis_title="Company
",yaxis_title="Number Of Vehicles ",title_x=0.5)
fig.show()
```

After executing the above code, the histogram is generated as shown in Fig. 9.3.

Figure 9.3 shows that Ford (447), Chevrolet, Mercedes-Benz, BMW, and Porsche are the top 5 auto manufacturers in Canada. These companies are followed by GMC, Toyota, Audi, Nissan, and Jeep. On further investigation, it is found that the popularity of vehicle manufacturers like Ford [1], Chevrolet [2], Mercedes-Benz [3], BMW [4], and Porsche [5] is not just because of their safety and handy features but also because of the high efficiency and lower carbon-emitting engines that these manufactures use.

B. **Vehicle Classes of the Cars**

The class of the vehicle affects the amount of fuel it consumes. A large SUV (sports utility vehicle) or a minivan will naturally consume more fuel resulting in more CO_2 emissions [6]. Therefore, it is necessary to study the popularity of different vehicle

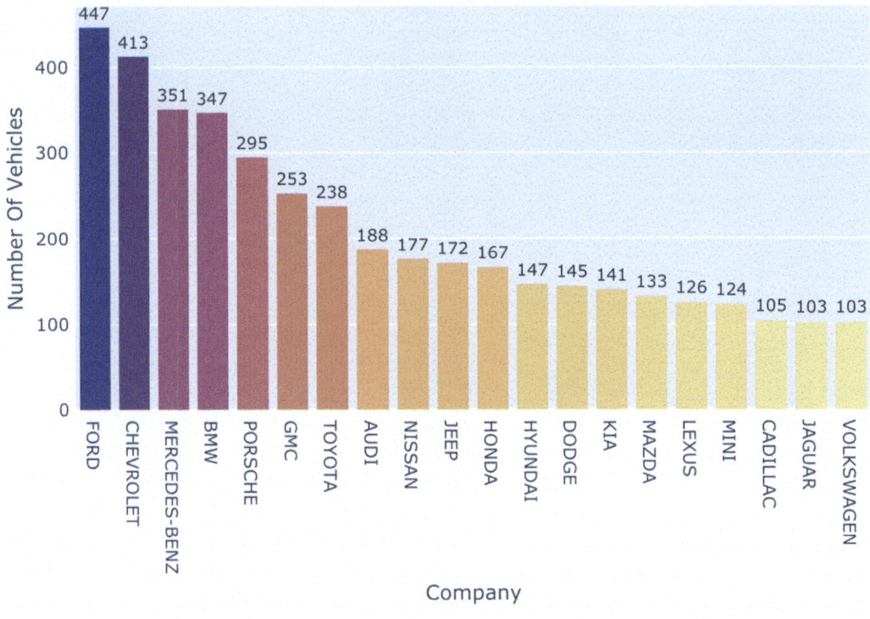

Fig. 9.3 The top 20 companies with the most cars manufactured in the last 5 years

classes. The code snippet mentioned below provides a visualization of the number of vehicles in each vehicle class:

```
vehicle_class = list(df['VEHICLE CLASS'])
for i in range(len(vehicle_class)):
vehicle_class[i] = vehicle_class[i].upper()
vehicle_class[i] = vehicle_class[i].replace(":", " -")
df['VEHICLE CLASS'] = vehicle_class
df_Vehicle_Class=df['VEHICLE CLASS'].value_counts().reset_index().
rename(columns={'index':'VEHICLE CLASS','VEHICLE CLASS':'Count'})
fig = go.Figure(go.Bar(
x=df_Vehicle_Class['VEHICLE CLASS'],y=df_Vehicle_Class['Count'],
marker={'color': df_Vehicle_Class['Count'],
'colorscale': 'Sunsetdark'},
text=df_Vehicle_Class['Count'],
textposition = "outside",
))
fig.update_layout(title_text=' VEHICLE CLASS',xaxis_title="VEHICLE
CLASS",yaxis_title="Number Of Vehicles ",title_x=0.5)
fig.show().
```

After executing the above code, the histogram is generated as shown in Fig. 9.4.

Figure 9.4 shows that Canada's most popular vehicle type is small-SUV, followed by mid-sized vehicles, compact vehicles, and standard SUVs. This could be mainly

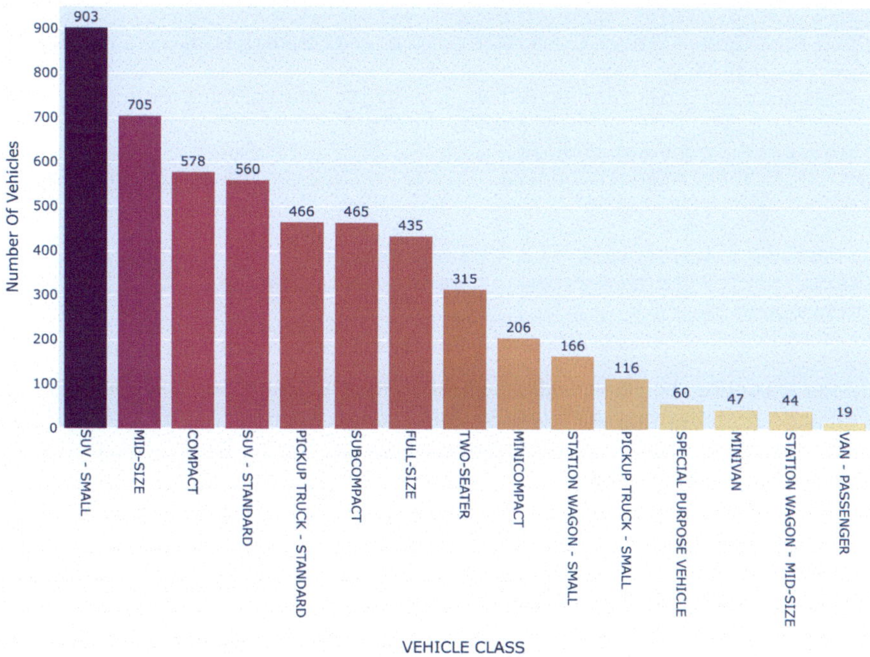

Fig. 9.4 Different vehicle classes provided by automakers in Canada

because of better fuel efficiency and lower carbon emissions. The vehicles like mini-vans and pickup trucks show lesser demand and popularity due to their higher fuel consumption.

C. **Most Popular Transmission Types in Cars**

Transmission types of a vehicle play an important role in the amount of carbon a vehicle releases, which makes it necessary to study its trends. Some of the abbreviations used for transmission types are as follows:

A = Automatic,

AM = Automated Manual,

AS = Automatic with Select shift,

AV = Continuously Variable,

M = Manual,

3–10 = Number of Gears.

The transmission types such as AS8, AS6, M6, and A9 are considered to be more efficient as it reduces fuel consumption resulting in less carbon emissions [7]. The code snippet discussed below gives a visualization of the number of vehicles in each transmission type:

```
df_Transmission=df['TRANSMISSION'].value_counts().reset_index().
rename(columns={'index':'TRANSMISSION','TRANSMISSION':'Count'})
fig = go.Figure(go.Bar(
x=df_Transmission['TRANSMISSION'],y=df_Transmission['Count'],
marker={'color': df_Transmission['Count'],
'colorscale': 'Sunsetdark'},
text=df_Transmission['Count'],
textposition = "outside",
))
fig.update_layout(title_text='Transmission Distribution
',xaxis_title="Transmission ",yaxis_title="Number Of Vehicles
",title_x=0.5)
fig.show()
```

After executing the above code, the histogram is generated as shown in Fig. 9.5.

Figure 9.5 shows that AS8 (1015) is the most common transmission system, followed by AS6 (660), M6 (533), A9 (394), and A8 (382), whereas transmission types such as AV10, AS5, A4, AM9, and AV1 are rarely found in vehicles. Hence, as mentioned above, it is observed that transmission types like AS8, AS6, M6, A9, and A8 are much more popular due to fuel efficiency and less carbon emissions.

D. **Vehicles with Different Types of Fuel**

Petroleum-based products are the most common reason for carbon emissions in the transport sector [6]. The different types of fuels used in vehicles are as follows:

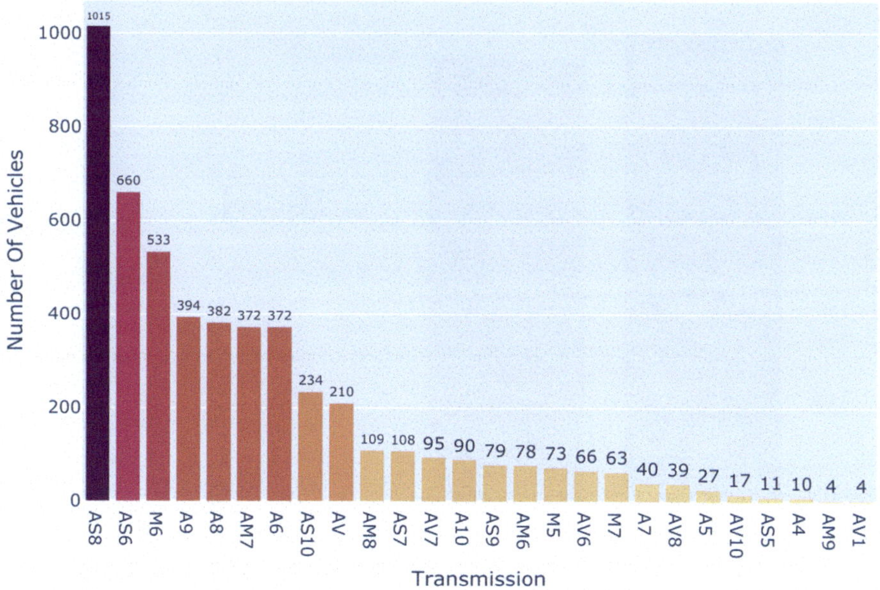

Fig. 9.5 Distribution of different transmission types in the vehicles

X is regular gasoline,

Z is premium gasoline,

D is diesel,

E is ethanol (E85),

N is natural gas.

The fragment of code mentioned below provides a relationship between the fuel type and the number of vehicles using each one of them.

```
df_Fuel_Type=df['FUEL TYPE'].value_counts().reset_index().renamex
rename(columns={'index':'TRANSMISSION','TRANSMISSION':'Count'})
fig = go.Figure(go.Bar(
x=df_Fuel_Type['FUEL TYPE'],y=df_Fuel_Type['Count'],
marker={'color': df_Fuel_Type['Count'],
'colorscale': 'Sunsetdark'},
text=df_Fuel_Type['Count'],
textposition = "outside",
))
fig.update_layout(title_text='Fuel          Type          Distribution
',xaxis_title="Fuel    Type    ",yaxis_title="Number    Of    Vehicles
",title_x=0.5)
fig.show()
```

After executing the above code, the histogram is generated as shown in Fig. 9.6.

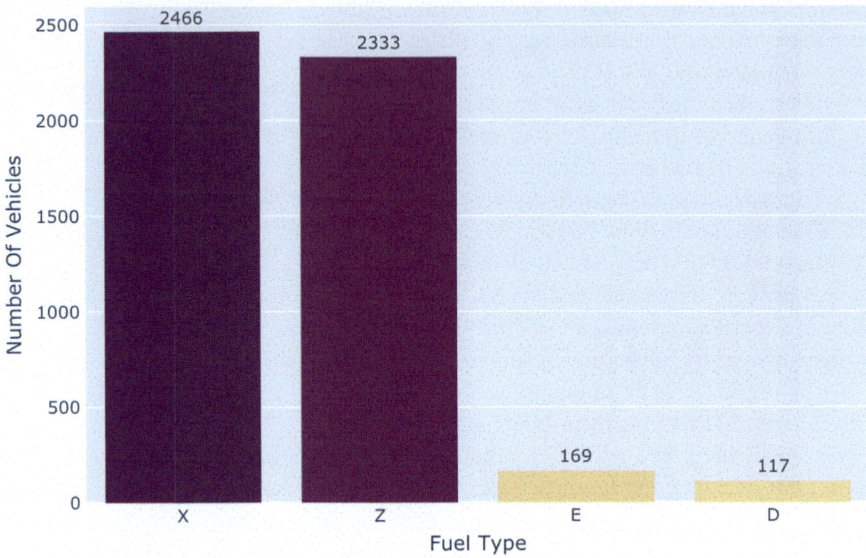

Fig. 9.6 Fuel types used in vehicles

Figure 9.6 clearly shows that most cars in Canada used X (2466) and Z (2333) as their fuel types, with E(169) and D(117) being on the lower spectrum. It can be concluded that gasoline is still the most popular fuel used in vehicles.

9.2.3 Correlation Analysis

In this section, correlations are studied to understand different relations between the variables. The study of correlations is important to discover relations between the dependent and the independent variables of the dataset. A scatter plot has been created to find such correlations, followed by Phik's correlation and Pearson's correlation.

A. **Scatter Plot**

Scatter plots between continuous variables and the target variable are plotted to find a correlation between the variables. The code snippet mentioned below shows correlation between "FUEL CONSUMPTION CITY (L/100 km (Liter/100 km))", "FUEL CONSUMPTION HWY (L/100 km (Liter/100 km))", "FUEL CONSUMPTION COMB (L/100 km (Liter/100 km))", "COMB (mpg—miles per gallon)" and "CO_2 EMISSIONS (g/km)":

```
from plotly.subplots import make_subplots
import plotly.graph_objects as go
```

```
columns = ['FUEL CONSUMPTION CITY (L/100 km)', 'FUEL CONSUMPTION HWY
(L/100 km)', 'FUEL CONSUMPTION COMB (L/100 km)', 'COMB (mpg)', 'CO2
EMISSIONS (g/km)']
fig = make_subplots(rows=2, cols=2)
fig.append_trace(go.Scatter(x = df[columns[4]], y = df[columns[0]],
mode="markers", showlegend = False), row=1, col=1)
fig.append_trace(go.Scatter(x = df[columns[4]], y = df[columns[1]],
mode="markers", showlegend = False), row=1, col=2)
fig.append_trace(go.Scatter(x = df[columns[4]], y = df[columns[2]],
mode="markers", showlegend = False), row=2, col=1)
fig.append_trace(go.Scatter(x = df[columns[4]], y = df[columns[3]],
mode="markers", showlegend = False), row=2, col=2)
fig.update_xaxes(title_text=columns[4], row=1, col=1)
fig.update_xaxes(title_text=columns[4], row=1, col=2)
fig.update_xaxes(title_text=columns[4], row=2, col=1)
fig.update_xaxes(title_text=columns[4], row=2, col=2)
fig.update_yaxes(title_text=columns[0], row=1, col=1)
fig.update_yaxes(title_text=columns[1], row=1, col=2)
fig.update_yaxes(title_text=columns[2], row=2, col=1)
fig.update_yaxes(title_text=columns[3], row=2, col=2)
fig.update_layout(height=1200, width=800, title_text="Scatter
Plots")
fig.show()
```

Figure 9.7 shows that all the continuous variables have a significant relationship with the target variable, i.e., CO_2 emission. FUEL CONSUMPTION CITY, FUEL CONSUMPTION HWY, and FUEL CONSUMPTION COMB show very similar positive relation with CO_2 emissions, which clearly indicates that with an increase in fuel consumption, there is an increase in carbon emissions. However, the relationship between CO_2 emissions and COMB (mileage/economy of the car) is negative, which means that higher mileage/economy corresponds to lower CO_2 emissions.

The profile report generated from the pandas profiling library creates an EDA report from which the correlation graphs are extracted.

```
import pandas_profiling
df.profile_report()
```

B. **Phik Correlation**

Phik (φk) is a new and practical correlation coefficient that works consistently between categorical, ordinal, and interval variables. It captures non-linear dependency and reverts to the Pearson correlation coefficient in the case of the bivariate normal input distribution.

Figure 9.8 shows the Phik correlation coefficient between all the variables in the dataset. It is observed that there is a high correlation between highway fuel consumption, city fuel consumption, combined fuel consumption, CO_2 rating, and CO_2 emissions. The high correlation between the independent variables can lead to the problem of multicollinearity, which can hamper the performance of the regression models.

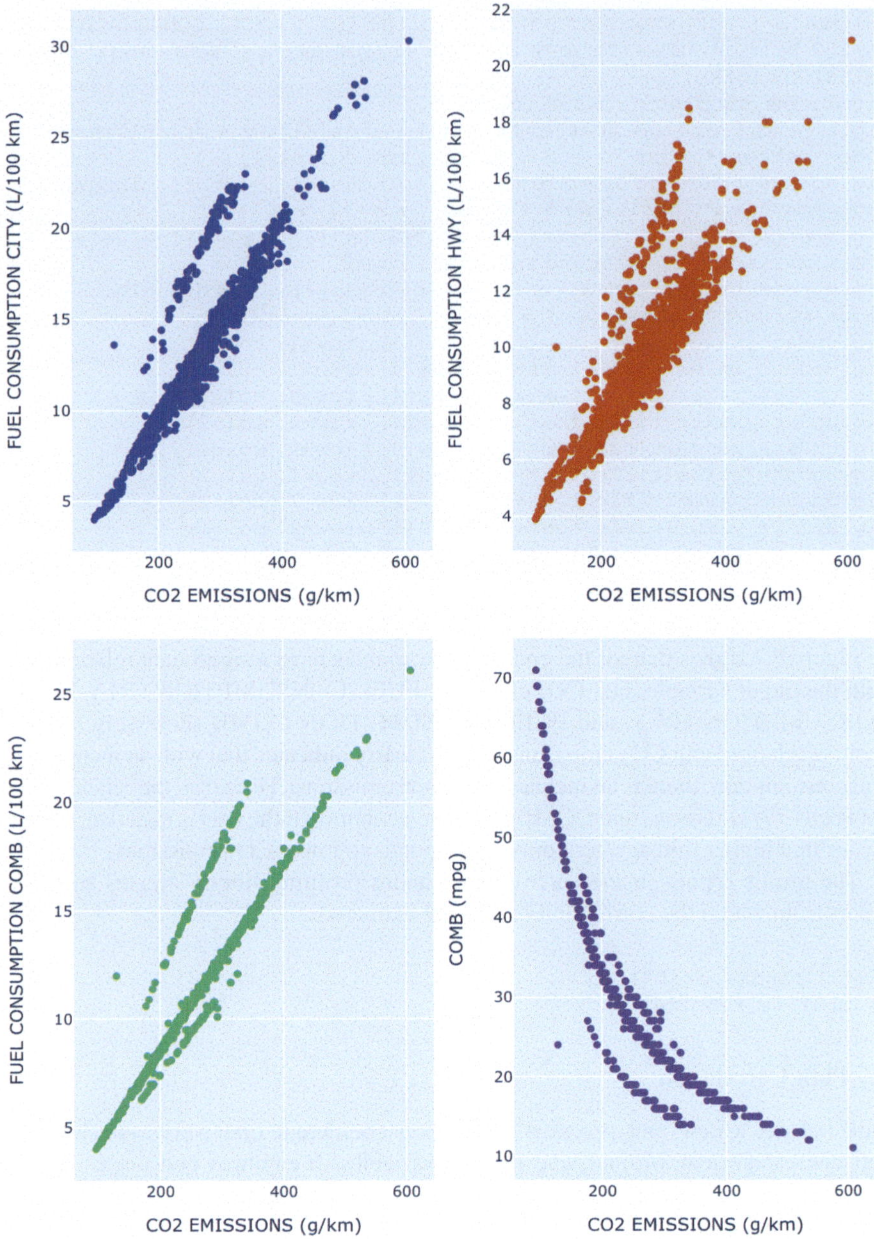

Fig. 9.7 Scatter plot between target variable and continuous variables

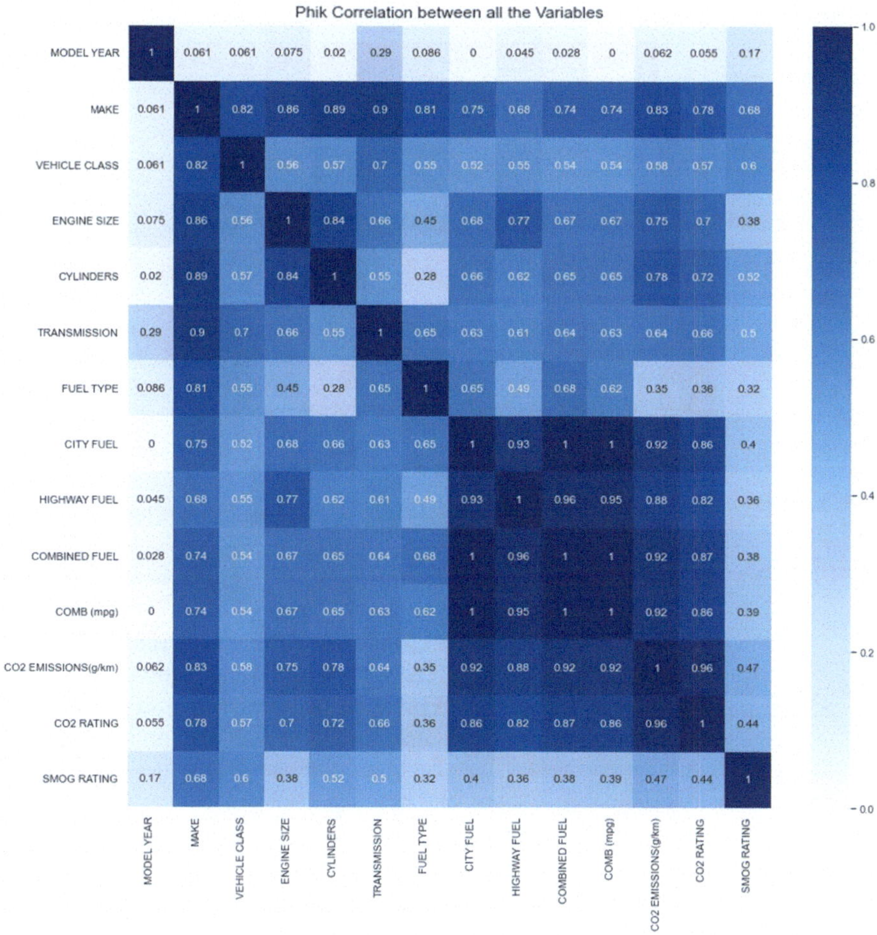

Fig. 9.8 Heatmap for Phik correlation

C. Pearson's Correlation

Pearson's correlation coefficient (r) is a measure of linear correlation between two variables. Its value lies between -1 and + 1, where -1 indicates a total negative linear correlation, 0 indicates no linear correlation, and + 1 indicates a total positive linear correlation.

Figure 9.9 shows Pearson's correlation between the continuous variables. Similar to Phik correlation (Fig. 9.8), it is observed that there is a high correlation between highway fuel consumption, city fuel consumption, combined fuel consumption, CO_2 rating, and CO_2 emissions. It is clear from Fig. 9.9 that COMB (mpg) is significantly negatively correlated (-0.95) with the target variable, i.e., CO_2 emissions (g/km). In contrast, the same target variable is significantly positively correlated

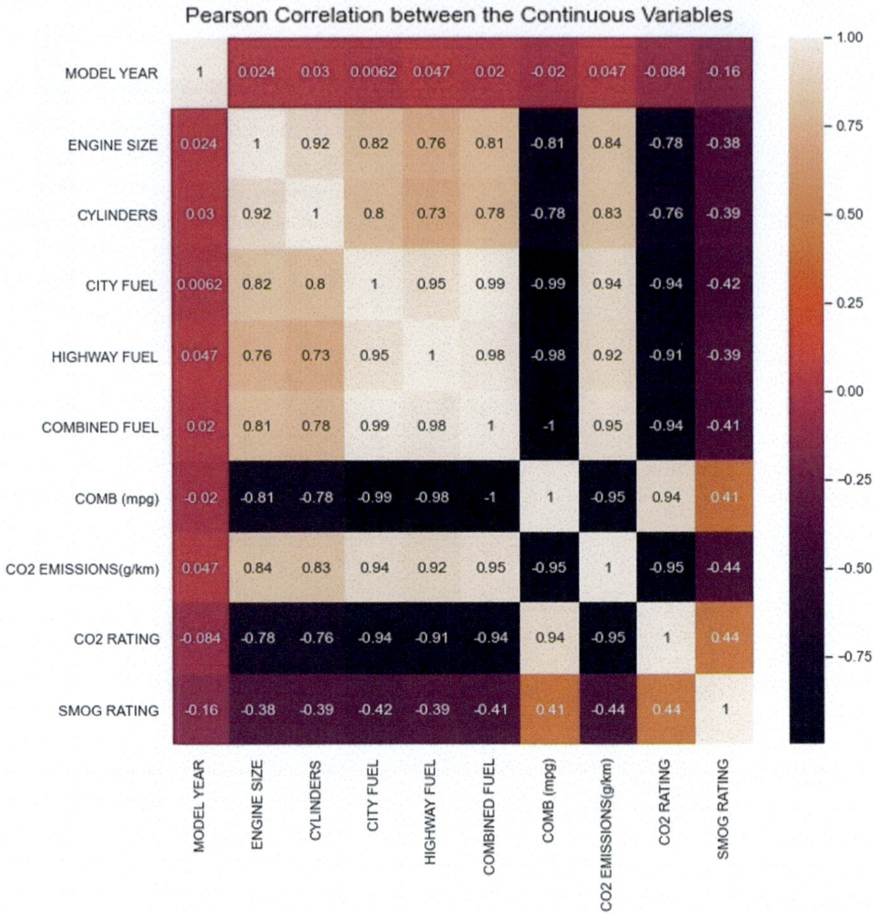

Fig. 9.9 Heatmap for pearson's correlation

with COMBINED FUEL (0.95), HIGHWAY FUEL (0.92), CITY FUEL (0.94), CYLINDERS (0.83), and ENGINE SIZE (0.84).

9.2.4 Feature Selection

The variables ("CITY FUEL" and "HIGHWAY FUEL") have been dropped to reduce multicollinearity. The variables ("MODEL YEAR" and "MAKE", "MODEL") have been dropped as these variables do not have any significant relation with the target variable. Following Python code is written to drop the variables from the final dataset:

```
df.drop(columns=['MODEL YEAR', 'MAKE', 'MODEL', 'CITY FUEL',
'HIGHWAY FUEL'], inplace = True)
```

9.2.5 Summary of Exploratory Data Analysis

In the first section of EDA, all the outliers are removed using the quantile transformer. In Sect. 9.2 of EDA, different features are analyzed to study their trends and impacts on carbon emissions. Section 9.3 deals with correlation analysis and feature selection. Now the dataset is processed, prepared, and ready for model building and evaluation.

9.3 Building a Model

In this section, six predictive models are developed, in which the machine learning models are trained on the dataset and also evaluated through their results. The explanation of these models is provided in Chap. 8. The model building pipeline is presented in Fig. 9.10. Next, data source identification, data extraction, data preprocessing, and EDA are carried out. After the EDA process, the categorical variables in the dataset are encoded in order to normalize. Two cases are considered for encoding the categorical variables in the dataset, i.e., frequency encoding and one-hot encoding. Normalization is important as it ensures that the entire dataset is on the same scale and hence prevents bias. Models are trained and evaluated over this dataset using K-fold cross-validation ($K = 10$), which helps to get a better idea about the performance of the models and ensures that every observation from the original dataset has the chance of appearing in the training and test set. Finally, all the results are compared, and the best model is selected for prediction. The evaluation metrics used are the R^2 score and RMSLE score (explained in Chap. 8).

9.3.1 Data Encoding and Data Normalization

Case 1: One-Hot Encoding

In one-hot encoding, each categorical value is converted into a new categorical column and a binary value of 1 or 0 is assigned to those columns. Each integer value is represented as a binary vector. One-hot encoding increases the dimensionality of the dataset. The categorical variables in the dataset (“TRANSMISSION”, “VEHICLE TYPE”, “FUEL TYPE”) are encoded using one-hot encoding, and the data is normalized using standard scaler. The dimensions of the dataset are (5085, 52). The fragment of code below is used to encode the categorical variables into one-hot encoding:

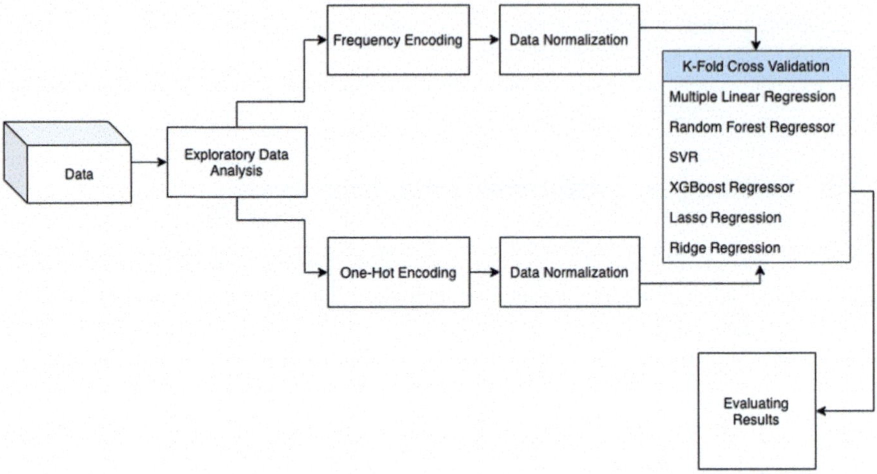

Fig. 9.10 Model building flowchart

```
df_transmission = pd.get_dummies(df.TRANSMISSION)
df_vehicle = pd.get_dummies(df['VEHICLE CLASS'])
df_fuel_type = pd.get_dummies(df['FUEL TYPE'])
df = pd.concat([df, df_transmission],axis=1)
df = pd.concat([df, df_vehicle], axis = 1)
df = pd.concat([df, df_fuel_type], axis = 1)
df.shape
(5085, 52)
```

The fragment of code below is used to normalize the data:

```
cols = list(df.columns)
from sklearn.preprocessing import StandardScaler
scaler = StandardScaler()
std_df = scaler.fit_transform(df.iloc[:,:51])
df.iloc[:, :51] = std_df
```

Case 2: Frequency Encoding

In frequency encoding, the frequencies of the categorical variables are calculated in a column, and then these frequencies are assigned to the categories in the respective column. Frequency encoding does not affect the dimensionality of the dataset. The categorical variables in the dataset ("TRANSMISSION", "VEHICLE TYPE", "FUEL TYPE") are encoded using frequency encoding, and the data is normalized using Standard Scaler. The dimensions of the dataset are (5085, 10). The fragment of code below is used to encode the categorical variables into frequency encoding:

```
def frequency_encoding(variable_frequency):
frequency = variable_frequency
keys = variable_frequency.keys()
```

```
variable_frequency = dict()
counter = 0
for i in keys:
variable_frequency[i] = frequency[counter]
counter += 1
return variable_frequency
transmission_frequency = df1.TRANSMISSION.value_counts()
trans_frequency = frequency_encoding(transmission_frequency)
vehicle_class_frequency = df1['VEHICLE CLASS'].value_counts()
veh_class_frequency = frequency_encoding(vehicle_class_frequency)
fuel_type_frequency = df1['FUEL TYPE'].value_counts()
ftype_frequency = frequency_encoding(fuel_type_frequency)
df1.TRANSMISSION = df1.TRANSMISSION.replace(trans_frequency)
df1['FUEL TYPE'] = df1['FUEL TYPE'].replace(ftype_frequency)
df1['VEHICLE CLASS'] = df1['VEHICLE CLASS'].replace(veh_class_
frequency) df1.shape
(5085, 10)
```

The fragment of code below is used to normalize the data:

```
cols = list(df1.columns)
scaler = StandardScaler()
std_df = scaler.fit_transform(df1.iloc[:,:9])
df1.iloc[:, :9] = std_df
```

9.3.2 Model Fitting and K-Fold Cross-Validation

The fragment of code below is used for training and evaluating the model using K-fold cross-validation (K = 10). The below code snippet also generates a small report for every model trained. The report includes the R^2 score, RMSLE score, and the training time for each iteration of the cross-validation, followed by the average R^2 score, average RMSLE score, and the average training time. This report is followed by a line plot between the training time and iterations for cross-validation.

```
from sklearn.inspection import permutation_importance
from collections import Counter
from sklearn import metrics
from sklearn.metrics import r2_score, mean_squared_error
from sklearn.model_selection import StratifiedKFold
from sklearn.ensemble import RandomForestRegressor, GradientBoost-
ingRegressor, AdaBoostRegressor
from sklearn.linear_model import LinearRegression, Ridge, Lasso
from sklearn.svm import SVR
from sklearn.model_selection import train_test_split
import time
import warnings
warnings.filterwarnings("ignore")
X = df.iloc[:, :51].values
Y = df.iloc[:, 51].values
```

```
def regression_evaluation(regressor, X, Y, regressor_name):
r2_arr = np.zeros(shape=(1,10))
rmsle_arr = np.zeros(shape=(1,10))
time_arr = []
cv = StratifiedKFold(n_splits=10, shuffle=True, random_state=42)
start_time = time.time()
for i, (train, test) in enumerate(cv.split(X, Y)):
itr_start_time = time.time()
regressor.fit(X[train], Y[train])
itr_end_time = time.time()
predict = regressor.predict(X[test])
r2 = r2_score(Y[test], predict)
rmsle = np.log(np.sqrt(mean_squared_error(Y[test],predict)))
r2_arr[0, i] = r2
rmsle_arr[0, i] = rmsle
train, test = train, test
time_arr.append(itr_end_time-itr_start_time)
print(f"Fold-{i}: R2_Score: {r2} RMSLE: {rmsle} time: {itr_end_time-
itr_start_time}")
end_time = time.time()
print()
print(f"Regression Report for {regressor_name}")
print("Average R2_Score :", np.mean(r2_arr))
print("Average RMSLE Score :", np.mean(rmsle_arr))
print("Total Time :", end_time - start_time)
fig = plt.figure(figsize=(15, 10))
ax.plot([0, 1], [0, 1], linestyle='--', lw=2, color='r',
label='Chance', alpha=.8)
plt.subplot(1,2,2)
plt.plot(time_arr, 'go-', color="darkblue")
plt.plot([max(time_arr) for i in range(10)], color='lightcoral',
label='UL')
plt.plot([min(time_arr) for i in range(10)], color='lightgreen',
label='LL')
plt.title("Training time for every iteration")
plt.xlabel("Iteration")
plt.ylabel("Time in sec.")
plt.legend(loc="upper right")
plt.show()
return np.mean(r2_arr), np.mean(rmsle_arr).
```

9.3.3 *Performance Evaluation*

After training the models, it is time to evaluate their performance. The performance evaluation of models built using the one-hot encoded dataset and frequency encoded dataset are shown in Tables 9.1 and 9.2, respectively.

It is observed from Table 9.1 and Fig. 9.11 that ensemble models, namely, XGBoost Regressor and Random Forest Regressor perform slightly better compared

Table 9.1 Performance evaluation of models trained on data encoded using one-hot encoding

Model name	R2 score	RMSLE
Multiple linear regression	0.9796	2.12
Random forest regressor	0.9980	0.74
SVR	0.9762	2.20
XGBoost regressor	0.9984	0.71
Lasso regression	0.9744	2.24
Ridge regression	0.9796	2.12

Table 9.2 Performance evaluation of models trained on data encoded using frequency encoding

Model name	R2 score	RMSLE
Multiple linear regression	0.9528	2.55
Random forest regressor	0.9981	0.74
SVR	0.9281	2.76
XGBoost regressor	0.9986	0.65
Lasso regression	0.9518	2.56
Ridge regression	0.9529	2.55

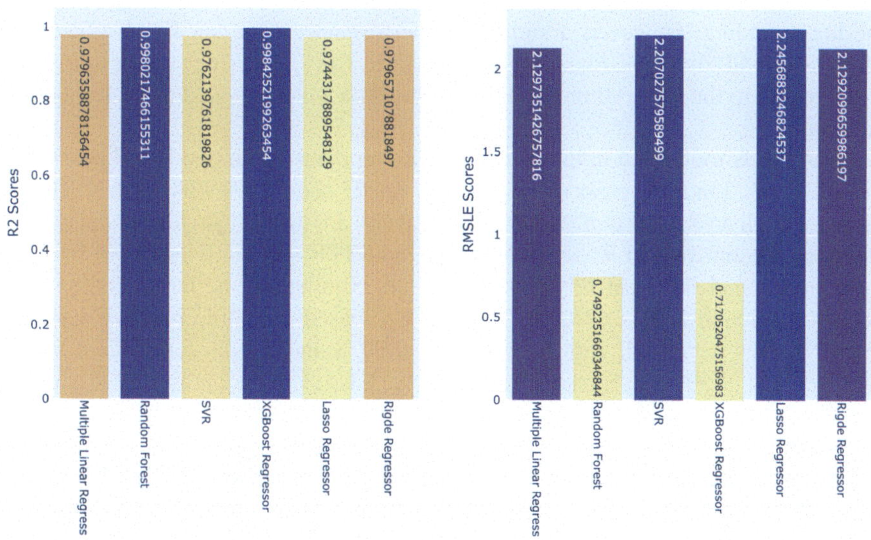

Fig. 9.11 Model comparison for one-hot encoding

to other models. These models show a high R^2 score and a low RMSLE, which is optimal for achieving good real-world performance.

It is observed from Table 9.2 and Fig. 9.12 that the performance of the ensemble models, namely, XGBoost regressor and random forest regressor, is the same as

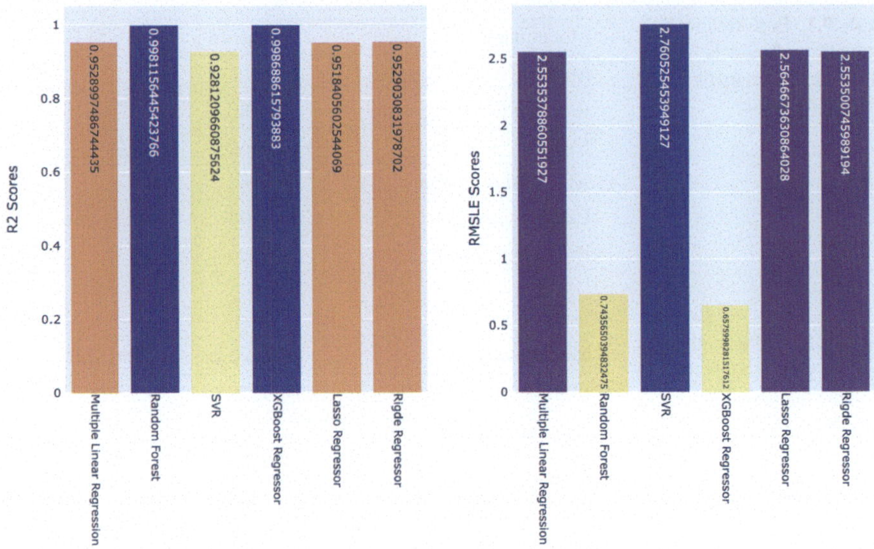

Fig. 9.12 Model comparison for frequency encoding

one-hot encoding. However, a significant performance loss can be observed for other models like multiple linear regression, SVR, Lasso, and ridge regression. This could be due to the lower dimensionality of the data when encoded using frequency encoding.

Finally, it is found that in the one-hot encoding dataset, the tree-based ensemble methods, viz., random forest regressor with an R^2 score of 0.9980 and XGBoost regressor with an R^2 score of 0.9984, respectively, are the best performers, whereas Lasso regression with an R^2 score of 0.9762 and SVR with an R^2 score of 0.9744, respectively, had the least R^2 score out of all. Similar to one-hot encoding, for frequency encoding, it is observed that the tree-based ensemble methods had the best R^2 scores out of all, with R^2 scores of 0.9978 and 0.9986 for random forest regressor and XGBoost regressor, respectively. Lasso regression and SVR had the least R^2 scores out of all the algorithms used, with R^2 scores of 0.9518 and 0.9281, respectively.

On comparing one-hot encoding with frequency encoding, it is perceived that one-hot encoding had an edge over frequency encoding. This is concluded because the machine learning models/algorithms performed slightly better in one-hot encoding than in frequency encoding. The probable reason behind the better performance might be the increased dimensionality of the dataset in one-hot encoding.

9.3.4 Predicting CO_2 Emission Using the Superior Models

According to the k-fold cross-validation (k = 10) results, random forest regressor and the XGBoost regressor are the best performers. Both the models were trained on Kaggle kernels without any GPU acceleration.

A. **Random Forest Regressor**

The random forest regressor gave an R^2 score of 0.9981 and an RMSLE score of 0.74 for one-hot encoded data, and an R^2 score of 0.9980 and an RMSLE score of 0.74 for frequency encoded data. The parameters set for this model were default parameters, where n_estimators = 100, min_sample_split = 2, and max_feautures = auto.

B. **XGBoost Regressor**

The XGBoost regressor gave an R^2 score of 0.9984 and an RMSLE score of 0.71 for the one-hot encoded data, and an R^2 score of 0.9986 and an RMSLE score of 0.65 for frequency encoded data. The parameters set for this model were default parameters, where booster (default trees) = gbtrees (gradient boosted trees), max_depth = 6, and ETA (learning rate) = 0.3.

In order to use these models for predicting CO_2, the pre-requisites are the information about the car such as “vehicle class”, “engine size”, “cylinders”, “transmission”, “fuel type”, “combined fuel consumption in the city”, “comb fuel in both city and highway in mpg”, “CO_2 rating”, and “smog rating”. The categorical variables like “transmission”, “vehicle class”, and “fuel type” can then be encoded by using one-hot encoding or frequency encoding, depending on the type of model. When inputs are fed to the model in the same order as mentioned above, the output of the model is the CO_2 emissions of the car. In order to scale up the prediction of CO_2 emissions by vehicles at an aggregate level, for example, at the city level, the model can take the inputs of all the cars in a consolidated manner.

9.4 Conclusions

The main objective of this chapter is to study the effect of different features of a vehicle on CO_2 emissions and to create a machine learning model that can accurately predict CO_2 emissions by any vehicle. After extensive data analysis, it is concluded that features like “FUEL CONSUMPTION CITY”, “FUEL CONSUMPTION HWY”, and “VEHICLE CLASS” have a direct effect on CO_2 emissions. As the fuel consumption on the highway and in the city increases or the vehicle size increases, the CO_2 emissions will also increase. On the other hand, a feature like COMB, basically the economy of the car, is inversely proportional to CO_2 emissions, i.e., better the economy, lower the CO_2 emissions. Thus, the focus should be given on to increase the economy of the vehicle which will help to achieve the net-zero emissions target.

The two superior models, namely, the random forest regressor and XGBoost regressor, can accurately predict the carbon emissions of any kind of vehicle that run on fossil fuels. These models can be used to predict carbon emissions caused by vehicles in metropolitan cities, which in turn can help the local governing bodies in regulating the rules and public transport systems, leading to a reduction in CO_2 emissions. In future, the current dataset, which is mainly about light motor vehicles, can be extended to include the data for other types of vehicles. The combined dataset can be used to create better insights regarding the CO_2 emissions and the type of vehicle responsible for it, and hence help make necessary changes or modifications in the manufacturing of those vehicles as well as policy formulation.

References

1. https://media.ford.com/content/fordmedia/fna/us/en/news/2020/06/24/ford-expands-climate-change-goals.html
2. https://media.chevrolet.com/media/us/en/chevrolet/home.detail.html/content/Pages/news/us/en/2015/nov/1118-carbon.html
3. https://www.mercedes-benz.com/en/vehicles/wltp/consumption-and-emissions/
4. https://www.bmw.com/en/innovation/sustainability-at-bmw.html
5. https://www.porsche.com/international/aboutporsche/e-performance/magazine/co2-neutrality-2030/
6. IPCC (2014) Climate change 2014: mitigation of climate change. Contribution of Working Group III to the Fifth Assessment Report of the Intergovernmental Panel on Climate Change [Edenhofer O, R Pichs-Madruga, Y Sokona, E Farahani, S Kadner, K Seyboth, A Adler, I Baum, S Brunner, P Eickemeier, B Kriemann, J Savolainen, S Schlömer, C von Stechow, T Zwickel, JC Minx (eds)]. Cambridge University Press, Cambridge, United Kingdom and New York, NY, USA
7. 5 Transmissions. National Research Council. (2015). Cost, effectiveness, and deployment of fuel economy technologies for light-duty vehicles. Washington, DC: The National Academies Press. https://doi.org/10.17226/21744

Chapter 10
Carbon Emissions Calculator: Impact of AI on Climate Change

10.1 Introduction

Global warming is one of the most critical issues on earth in modern days. Many research works are carried out that caution towards the increase in global warming above pre-industrial levels, which could adversely impact the entire human existence on earth, if necessary actions are not taken. Carbon emissions from industries, houses, shops, temples, forests, etc. have resulted in higher levels of CO_2 in the atmosphere. CO_2 is being consumed majorly by the flora, but due to reckless deforestation and slow reforestation, the amount of CO_2 is continuously increasing. This gas is considered to be a major component of Greenhouse gases (GHGs) which points to global warming and eventually leads to climate change.

It would be prudent to look at the effect of technology on climate change, and the most popular technology of current times is Artificial Intelligence (AI). However, AI has been with humankind since the very existence of computer systems. The evolution of human creativity and meticulousness has contributed to the progress in AI systems. For example, people used books to store information from earlier times and search through them when required. Nowadays, almost all information is stored in the computer system which has made searching very easy. Back from the days of 1957, AI has made its prominent place in society and is burgeoning to garner more importance. Despite such admirations, AI had to face certain obstacles along the way. One of the hindrances posed was the lack of computational power of AI systems. Another setback was due to the decreasing amounts of available funds to carry out research and development initiatives. However, facing this challenging time, researchers worked on the flaws that occluded the path of AI from making lives more manageable. Several goals were set, many have gotten to reality, and some remained in books and papers only. The Japanese Fifth-Generation project could be an example of failure in attempting to advance the technology [1]. The project was ultimately put to an end after investing a total of 400 million USD on it. This is such a huge loss to the economy and the resources that were used in the project. In this

N. Sharma and P. K. De, *Towards Net-Zero Targets*, Advances in Sustainability Science and Technology, https://doi.org/10.1007/978-981-19-5244-9_10

way, with trial and error, often with calculated mindsets, humans have advanced the scenario for AI systems and achieved success in many places.

With time, the computational power began to grow to such an extent that it started growing exponentially. Petabytes, exabytes, and zettabytes are common terms among many industries nowadays. A strong subcategory of AI evolved with data, i.e., Machine Learning (ML). Systems can now imitate human behavior and make decisions on their own by only learning the patterns in the data, just as a child does when it starts learning things. One crucial example of this could be IBM's Deep Blue, a chess playing program, which has the capabilities to compete with any human, and it even defeated the world chess champion Gary Kasparov [2]. While there is still much debate on the initiation of a neural network, a study by Fukushima and Sei is considered as the starting place of neural networks in computational systems [3]. Many innovations then followed, and AI has been utilized in various sectors ranging from normal diagnostic tests to explore secrets of the Universe. These systems require tremendous amounts of computation which can be accomplished by consuming energy and making systems more efficient. This has been achieved by the exploitation of nature. Mining and clearing of lands for huge power grid stations are some of the examples of it.

Since everything comes with a price, and so does AI. Advancement in AI has started showing its effect on the environment. To power AI systems, energy is required, which comes from electricity. Most of the electricity comes from non-renewable sources of energy, which in turn could lead to more carbon dissipation. According to a study by Dhar, the contribution of AI to climate change has increased exponentially in recent decades [4]. An increase in computation far exceeds what is predicted by Moore's law. Despite pledging to net-zero carbon emissions, several companies increased their carbon footprint. For instance, a study by the University of Massachusetts Amherst shows that training a single big language model equates to about 300,000 kg of carbon dioxide (CO_2) emissions. Looking over this exhaustion of resources for just a small increase in performance, scientists across the globe proposed various measures to mitigate the contributions of AI to climate change [4]. Lacoste et al. proposed a machine learning emissions calculator, which is used to quantify the amount of carbon dioxide (CO_2) emissions on training the model [5]. Moreover, several researchers now focus on considering efficiency as a factor when looking over approaches proposed in AI conferences. Scientists now realize that the negative contribution of AI to climate change is increasing at a very fast pace, and an exigent shift towards Green AI has become inevitable [6]

This chapter aims to make a point that not only positive but negative actions are also performed by AI, which deteriorates the environment to some extent. The inconsistencies between different processors and cloud providers are also studied and demonstrated using various visualizations. Furthermore, this chapter proposes a real-time carbon emissions calculator for personal computers, which would help to find out the system's exploitation in terms of its carbon emissions dynamically. The remaining part of the chapter is organized as follows: Sect. 10.2 takes into account the literature review, which is further bifurcated into a contemporary and systematic literature review. Section 10.3 deals with the different types of AI, their applications and implications on the environment. While Sect. 10.4 presents the study of various real-time carbon emissions calculators, Sect. 10.5 discusses the materials and methods used for the experiment. Section 10.6 presents experiments, the ML model, and the results of the analysis. Finally, Sect. 10.7 provides future steps in the direction of this study, along with conclusions.

10.2 Literature Review

The surprising relationship between AI and climate change has always intrigued the community. As the levels of global warming increase, attempts towards ameliorating the conditions have also been made, and AI has been found to have a supporting hand in it. However, some studies also focused on the complications that AI introduces to the environment. The following section is subdivided into two parts: contemporary literature review and systematic literature review. Contemporary literature review deals with having a detailed analysis of the previous studies performed in the particular domain. On the other hand, a systematic literature review presents the analysis of the importance of the literature and the relevance of the topic. Using statistical and non-statistical approaches, one discusses the potential scope for the discussion of the topic under study. A survey using both approaches is necessary to understand the progress in the subject in a clear and precise way.

10.2.1 Contemporary Literature Review

In the recent decade, there has been a rigorous study on the relation between AI and climate change. Some studies focused on calculating the carbon emissions from training ML models. The study by Dhar exposes the overuse and wastage of computational power in training giant ML models [4]. Moreover, the study also suggests using pre-trained models rather than implementing the same from scratch in order to extenuate the implications of carbon emissions. It is very important to make global warming fettered under 1.5 °C above pre-industrial levels, as has been reported by IPCC [7]. Several approaches to calculate carbon emissions of ML models have been proposed. For instance, the study by Lacoste et al. proposes an ML Emissions Calculator and the factors imperative for it [5]. The study put forth two factors, namely, (a) type of energy used and (b) computing infrastructure and training time on which CO_2 emissions depend. The dataset for the same is also available on the GitHub repository [8]. Although the study has found a way to calculate carbon emissions of ML models, it does not claim its preciseness in the dearth of adequate information. While a way to calculate the carbon emissions of AI systems is found, there is a need to be wary about the critical research on developing these AI systems. Schwartz et al. introduce two types of AI, viz., "Green AI" and "Red AI" [6]. The Red AI is formulated as the AI research which focuses on getting novel results with enormous computational cost. At the same time, Green AI is the research that focuses on getting a novel result by using efficient tactics and with minimal computation cost. It also gives an equation to calculate the cost of producing a result in ML systems. Floating Point Operation (FPO) has been chosen as an efficiency measure in the study. Furthermore, Strubell et al. infer on some predominant Natural Language Processing (NLP)-based ML models, i.e., transformer, ELMo, BERT, and GPT-2, by calculating the environmental and financial cost of fine-tuning these architectures [9]. The financial costs are found by calculating the kilowatt hour (KWH) energy consumed in training such models. The research also provides actionable steps to dampen the contribution of NLP models to climate change. Public awareness about climate change does not meet the anticipated level [10]. This could be due to the fact that people are more readily able to trust things, which they encounter or can visualize. Following this thought process, the study by Schmidt et al. proposed a CyclicGAN model which has been utilized to plot flood images of houses and buildings to visualize the possible impact of floods in particular areas [11]. A distended version of this project is intended to include the decision knobs and image plotting for other disasters such as volcanoes, earthquakes, etc. The study by Rolnick et al. discusses the different steps taken by the ML community to address the problem of climate change [12]. They have shown the positive contributions of ML in various domains such as electrical systems, transportation, building and cities, farms and forests, etc. The study by Patterson et al. provides the intel about energy usage and carbon footprint of training large energy models such as T5, Meena, GShard, Switch Transformer, and GPT-3 [13]. The study also put forth the formula to calculate the carbon footprint. Durgam et al., in their research, talked about the effects of radiation on the life of fauna (especially birds) and presented the results of their survey on the mobile towers in the Bijapur District affecting the number of birds in the respective areas [14]. Table 10.1 recapitulates the key information extracted from the literature being studied.

From the table, it can be inferred that the various approaches used to fight against the issue of climate change and most of them also discuss the negative instances of implementing AI models inefficiently. These approaches have been introduced to mitigate the proliferation of the AI community in affecting the environment.

10.2.2 Systematic Literature Review

This section deals with the systematic survey of the literature, which is similar to the domain of climate change and AI. The data has been collected from IEEE Xplore using the advanced search with keywords "carbon footprint", "carbon emissions", "green AI", "CO_2 emissions", "greenhouse gas emissions", and "carbon tax". A union of the results from these keywords has been included in the study.

Figure 10.1 shows a spider chart elucidating the number of research (conferences, journals, books, etc.) published in the period spanning from the year 2011 to 2021. An increasing trend can be observed from the chart, with the highest number of publications (1032) in the year 2019. An eventual growth of research focus in this

Table 10.1 Contemporary literature review

Title	Research objective	Dataset utilized	Highlights	Application
Carbon impact of artificial intelligence [4]	1. Introducing exploitation of resources for AI systems and ways to mitigate it	–	Red AI, Green AI, NAS, Carbon Footprint, Global Climate Strike, Moore's Law, AutoML	Preliminary study of the relation between AI and Climate Change
Quantifying carbon emissions of machine learning [5]	1. Introduce an ML carbon emissions calculator	Data collected from various sources majorly till 2019	Power Usage Effectiveness (PUE), Renewable Energy Certificates (REC), Thermal Design Power (TDP), Carbon Neutral, Carbon offsetting	Calculate carbon emissions of machine learning models
Green AI [6]	1. Efficiency as essential evaluation criteria 2. Add price tags to various ML models 3. Formula to calculate the cost of training	–	Green AI, Red AI, FPO, NAS	ML model performance evaluation using FPO
Visualizing the consequences of climate change using cycle consistent adversarial networks [11]	1. A system to visualize the impact of floods in various places	1. Data collected from open-source websites 2. Images have been taken before and after the flood 3. Suburban detached houses, urban townhouses, and apartment buildings are targeted	CyclicGAN, climate science, climate model	Climate change mitigation systems

(continued)

Table 10.1 (continued)

Title	Research objective	Dataset utilized	Highlights	Application
Tackling Climate Change with Machine Learning [12]	1. A comprehensive study on the positive contributions of AI in different domains	Data for several domains have been collected separately	Solar fuels, distribution grid, nuclear fission, electricity curtailment	Applying Machine Learning for mitigating the consequences of climate change
Energy and Policy Considerations of modern deep learning research [13]	1. Calculate the cost of training and inferencing of NLP models as per the proposed method 2. Actionable steps to mitigate climate change	Data has been collected from different online sources such as the New York Times, the University of Michigan report on Carbon Footprint, etc	NAS, transformer, BERT, GPT-3, ELMo, BLEU score	The study focusing the proliferation of Red AI
Effects of Mobile Tower Radiation on Birds in Bijapur District, Chhattisgarh [14]	1. Study of adverse implications of electromagnetic radiations on fauna	The survey has been taken in the available nine mobile tower areas across Bijapur. The inference was made 2–3 h in the morning and evening	Electromagnetic radiation, mobile tower, radio frequency	Study on systems to mitigate the effect of electromagnetic radiations on fauna

area is clearly recorded, which implies the importance of this research topic. It may be noted here that the data for the year 2021 is not for the entire calendar year.

Figure 10.2 shows the pattern of the type of publication and their count in the respective years. This chart helps us zoom into the publications and infer about them. It is visible that a maximum number of contents are presented in conferences, whereas publications in the form of books are minimal throughout the period. This implies that the topic is not mature enough and still evolving, and therefore there is enough scope for further research in this domain.

Figure 10.3 describes a word cloud of the keywords extracted from the research papers being studied for the systematic literature review. Out of these keywords, *field_emission*, *carbon_nanotube*, and *carbon_emission* are the ones which are most widely used in the literature. This word cloud provides a direction to the researchers in their study by highlighting the frequently used key terms in previous studies.

After a thorough review of literature, in both contemporary and systematic ways, it is evident that AI comes with lots of advantages and side effects too. The objectives of this chapter are mentioned below:

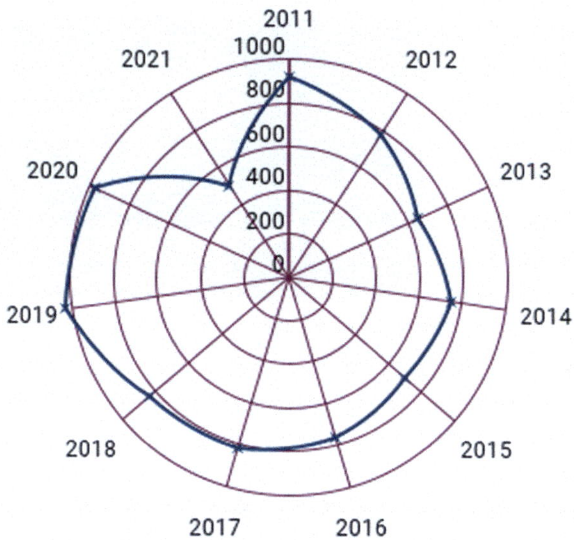

Fig. 10.1 Radar chart showing the distribution of publications from 2011 to 2021

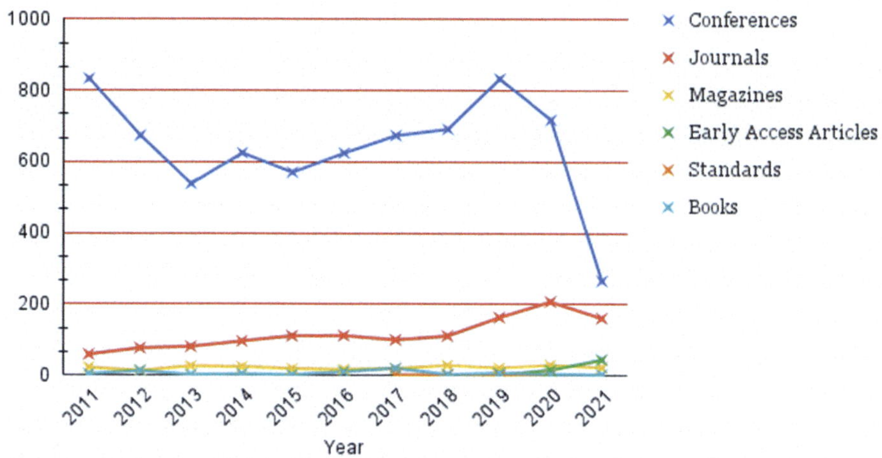

Fig. 10.2 Line plot showing the type and number of publications from 2011 to 2021

1. To understand various types of AI and its efficient applications.
2. To study the implications of effective use of AI and ways to keep its adverse effects on climate change at a minimum.
3. To analyze the different combinations of hardware, region of computing, and cloud providers for machine learning models to get a glimpse of their carbon footprint.

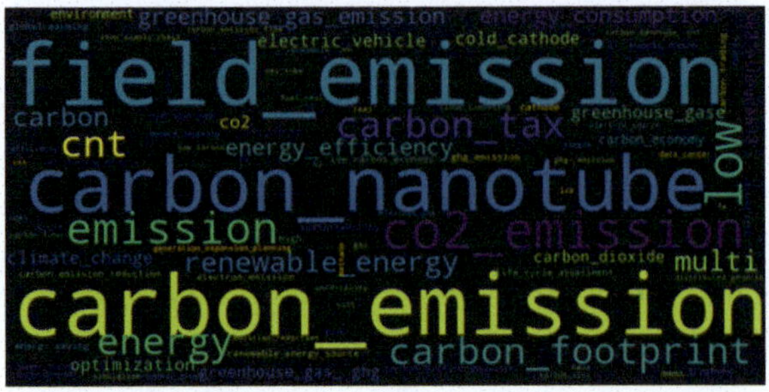

Fig. 10.3 Word Cloud of the most common words used in the research studies

4. To discuss various real-time carbon emissions calculators and build a real-time carbon emissions calculator for Personal Computers (PC).

10.3 Types of AI, Its Applications and Implications

Along with all the motley of areas and approaches covered in this book which replicates a good stature of AI in terms of achieving net zero, there are some areas for improvement for AI towards climate change. Whenever fossil fuels are burnt, they release a massive amount of CO_2, which is one of the prominent components of GHGs and also a significant contributor to global warming. Despite the consensus by 196 countries at the 21st Conference of Parties of the United Nations Framework Convention on Climate Change (also termed as the Paris agreement) to limit carbon emissions to 1.5 °C above pre-industrial levels by 2030, minimal efforts are seen to achieve such a goal [7]. For instance, regardless of pledging to net-zero carbon emissions by 2040, Amazon's carbon footprint has increased by 19% in the last year [15]. Large ML systems are also a major concern for climate change since training and inferring these systems require a lot of computational power, which, in turn, consumes an equivalent amount of electricity. This electricity is, in order to avoid costs, usually generated from burning non-renewable sources of energy since they are comparatively cheaper than installing and maintaining renewable sources of energy. On the basis of efficient usage of resources and computational complexity, AI has been bifurcated into two types, namely, Green AI and Red AI.

10.3.1 Green AI

Research that solicits on obtaining novel results without increasing computational cost, and, in most cases, reducing it is being termed as Green AI. While considering Green AI, efficiency and not accuracy are taken into account for evaluating a model's performance. A few of the conferences nowadays have started focusing on efficiency as a measure to select models. The study by Schwartz et al. provides us with information about Green AI and the need for its burgeoning popularity [6].

10.3.2 Red AI

Research that solicits on obtaining avant-garde accuracy on account of massive computational cost is termed as Red AI. With the advent of industrialization, the competition among researchers and companies to provide the best possible service or product has started to advance. People invest a huge sum of money in improving the accuracy of ML models, even by 0.1%. The Floating Point Operation (FPO) cost of training such models has increased dramatically during the recent decades while having a much smaller increasing curve for the accuracy. In the following part of this section, how AI reflects as a challenge to climate change is discussed:

A. **Competition to Thrive**

In recent decades, the technical community has seen an exponential increase in the research and development of various products and services. These entities are being used at an enterprise level, which eventually leads to a more competitive environment and zeal to achieve maximum performance at almost any cost. There are various cases where immense computational power has been utilized to increase the accuracy of the system even by a negligible amount. For instance, consider a company that has launched an innovative product, which can trigger the rivals to introduce a similar product with the same or slightly improved features, which could be an extensive usage of resources for redundant purposes.

B. **Neural Architecture Search**

Training every ML model requires the hyper-parameters to be well optimized so as to reach global minima. Choosing the wrong set of hyper-parameters could lead to training incessantly or poor performance. This is the reason why hyper-parameter optimization techniques are used. However, according to a study by Bergstra et al., a particular dataset is responsive to only a specific set of hyper-parameters, and the remaining other hyper-parameters clearly do not matter much, which makes grid search (a greedy technique for hyper-parameter search) an inferior choice than a random search [16]. Hence, Neural Architecture Search (NAS) techniques are needed to be optimized urgently. For example, in a 2019 report by So et al., NAS acquires a BLEU (Bilingual Evaluation Understudy) score of 29.7 for English to German translation [17]. It is important to highlight that the score is improved by 0.1 from the previous one at the cost of 150 thousand USD [17].

C. **Models with Non-efficient Parameters**

A review of an ML model, i.e., Transformer, is carried out in this section. This model has achieved admirable accuracy but utilizes tremendous computational power and hence has become less efficient. A transformer is an encoder-decoder architecture with several self-attention heads, which helps it to accomplish tasks like machine translation very easily. The number of parameters in a typical transformer model could range from 65 to 213 M. However, recent discoveries with NAS have brought a new type of Transformer, i.e., "Evolved Transformer", which outperforms the vanilla transformer model by a BLEU score of 0.7 [17]. The Evolved Transformer showed a decrease in the net CO_2 emissions for model training in metric ton from 0.1357 (Transformer-Big) to 0.01777 (Evolved Transformer-Medium). It has also saved 48.5 tons of CO_2 emissions while training the Meena DNN compared to the naive Transformer.

D. **Electric Vehicles**

Electric Vehicles (EVs) have become the new generation of automobiles replacing the previous internal combustion-based engines. The majority of energy in EVs is through electricity, which is being fetched from batteries. These batteries are not ordinary ones but highly robust and efficient ones. Since EVs work with electricity from batteries, it has minimized the use of non-renewable sources of energy for the same purpose. Thus, it is, in some way, contributing to Green AI [18].

What if the electricity at the charging station comes from non-renewable sources of energy? It could also be possible that the electricity which comes from non-renewable sources of energy is being generated through inefficient means, and this could release more CO_2 into the environment. One more thing to note is that EVs are powered by batteries with extremely long retention power, as they are made up of Rare Earth Elements (REE) such as cobalt, nickel, lithium, etc. These elements are stored deep into the earth's surface, and thus, it is required to be mined, leading to an eventual increase in the carbon footprint [19]. Therefore, EVs are not completely green.

In addition, EVs are also embedded with lots of AI capabilities to discover high-performing and efficient batteries provide realistic driving-range estimation and suggest an optimal association with grid technologies. Nevertheless, in order to improve the performance of EVs, the companies are making wide applications of AI, which may lead to Red AI if not used judiciously [20].

Another aspect to think about is that the majority of the batteries used in electric vehicles are non-disposable. According to a study by the International Council of Clean Transportation (ICCT), 99% of lead-acid batteries are being recycled. In contrast, most of the batteries powering electric vehicles, such as lithium-ion batteries, are either burnt down or thrown in landfills [19].

E. **Information and Communication Technologies, and Digital Electronic Systems**

AI is not only confined to ML but also includes other forms of technologies, which operate on rules and are termed as rule-based AI. In this section, a few of the environment-related challenges faced due to rule-based AI systems used in IoT devices, smart watches, smart phones, computer software, cameras, e-readers, flash drives, television, communication devices, etc. are addressed. These systems directly or indirectly have a debilitating effect on the climate and, thus, contribute negatively to its change.

Mining is a major concern caused indirectly by electronic devices to the environment. Since mining is used for various purposes, most of it is done to prepare these electronic devices. Another aspect which vitiates the climate is the disposal of these devices. While some are easily disposable and even recycled, others need special processing, and a few of them can never be disposed of. For example, batteries that are non-rechargeable or single use, such as AAA and AA, which are common in flashlights, toy cars, remotes, clocks etc., are not easily disposable and pose a serious threat to the environment.

In this era of global connectivity, people have developed ways to connect to anyone around the globe using wireless means. Several wireless devices such as drones, mobile towers, smartphones, self-driving cars, and other IoT devices are really crucial for the advancement of humankind but have some undesirable effects on the environment. The radiation emitted from such devices causes harm not only to our health but also to the environment and other living beings. Many birds navigate using the natural electromagnetic (EM) radiations, but these introduced artificial EM radiations have caused birds to disorient from their normal pathway, get lost, and eventually die [21]. Since birds are a part of the ecosystem, there would be an adverse effect on humankind too, if their species became extinct.

Wireless technology is evolving at an admirably fast pace. With the advent of 5G (fifth generation) technology, directing the attention of many people towards its deteriorating effects on the environment has become an imperative task. The plan of the 5G network is to make data processing much faster.

Consider an analogy by Nordrum et al. that 5G would allow to download an HD film in a fraction of a second, which took about 10 min on a 4G network. Therefore, 5G will provide unprecedented speed, which, when coupled with AI, will offer a different experience to the users. This great service inherently targets the relentless and exhaustive use of natural resources. In order to set up a fully functional 5G network, five different components need to be installed, i.e., millimeter waves, small cells, massive MIMO, beamforming, and full duplex [22]. These devices are deemed to be implemented on a wider level than 4G networks which ultimately would feed up on a lot of energy and expose to greater amounts of radiation which would be detrimental to the environment.

F. AI-Powered Social Media Energy Usage

Wide usage of social media applications is seen these days which is largely powered by AI. This has been possible majorly due to the following reasons:

Internet Availability

With the advent of efficient technologies to generate and maintain the Internet, people are provided with this benefit at cheaper rates. Moreover, an extensive range of products is introduced nowadays which cannot be operated without the Internet and thus having an Internet connection has become a necessity for everyone. These products also comprise of social media applications such as Facebook, Instagram, WhatsApp, Twitter, etc.

Global Connectivity

Social Media has been considered as the best approach to connect with people from any corner of the world. Not only personal but professional engagements are also being handled using these platforms. The mailing system is also an extensively used message sharing platform. It is estimated that a normal email with no graphics involved takes up to 4 g of CO_2 emissions, and an average individual releases up to 135 kg of CO_2 by sending mails [23]. Since people don't see the inherent cost of sending emails, they keep on unnecessarily forwarding them and increasing their carbon footprint.

This extensive usage of social media requires a lot of research, development, and maintenance consuming huge amounts of electric power, which majorly comes from non-renewable sources of energy. A 2017 report by *Clicking Clean* elucidates the consumption of energy by different platforms such as Facebook, Google, Netflix, HBO, etc. [24].

From Table 10.2, it is clear that social media platforms such as Skype, LinkedIn, and Twitter needed immediate attention as their power extraction from non-renewable sources of energy exceeded the same from renewable sources of energy [24]. Among these, LinkedIn and Twitter had clean energy consumption of only 10% out of their total energy usage. These are such platforms which are hugely used by people, and thus their need to become more and more green is a dire necessity.

Table 10.3 provides insights about energy usage bifurcation of companies with prime business around search engine and recommendation systems. It is evident that Apple is the greenest company (according to the 2017 report), followed by Google, HBO, and Netflix [24]. Again, Netflix is the most popular video streaming service available, and thus it has the opportunity to move towards becoming an absolute green company.

After studying the various types of AI, their different applications and their implications in this section, it is clear that the inappropriate use of AI might have side effects on the climate. Hence, the carbon emissions calculator can be useful to keep a check on the judicious use of AI.

Table 10.2 Energy usage of different social media platforms (in percent)

Company	Clean energy	Natural gas	Coal	Nuclear energy	Others
Facebook	67	7	15	9	2
Spotify	56	15	14	10	5
Skype	32	23	31	10	4
LinkedIn	10	31	23	20	16
Twitter	10	43	21	14	12

Table 10.3 Energy usage of search engine and recommendation-based systems (in percent)

Company	Clean energy	Natural gas	Coal	Nuclear energy	Others
Google	56	14	15	10	5
Netflix	17	24	30	26	3
HBO	22	20	25	25	8
Apple	83	4	5	5	3

10.4 A Study of Existing Carbon Emissions Calculators

There are various approaches used by researchers to calculate the carbon emissions of the system. In this section, some of the popular carbon emission calculators are discussed.

10.4.1 Machine Learning Carbon Emissions Calculator ($MLCO_2$)

$MLCO_2$ is a tool to calculate the carbon emissions of ML models, which has been developed by Alexandre Lacoste et al. [5]. In their study, the authors used CO_2e to calculate the carbon emissions, which takes the following details as inputs for the calculator:

A. **Hardware Type**

Since ML systems can be implemented on a variegated set of hardware and whose correct choice is very crucial for the training and inference of the model, the calculator includes different options of processing systems such as Intel Xeon E5, GTX 1080 Ti, AMD RX480, etc.

B. **Hours Used**

Different ML models take up separate time for training on a distinct set of hyper-parameters. These criteria are thus included in the calculations of carbon emissions.

C. **Cloud Provider**

Every cloud provider has its way of implementing the system and getting energy for it. This energy could be mostly from renewable sources, which is considered a best practice. However, a notable part can also be from non-renewable energy sources. Thus, information about cloud providers and their carbon emissions is also very imperative before training the model.

D. **Region of Compute**

Exhaustible and inexhaustible sources of energy are being distributed on the basis of regions, and thus having information about the zone in which the model is being trained is very critical.

10.4.2 Carbon Footprint Calculator, the Nature Conservancy

Apart from the carbon emission calculator for extracting the carbon footprint of ML systems, calculators for everyday tasks such as travel, home, food, shopping, etc. have also been proposed [25, 26]. This calculator calculates the individual carbon footprint and is only functional for US states and counties. It generates three types of outputs, as mentioned below:

A. **Your {Activity} Footprint**

This output shows the carbon footprint for one of the activities mentioned above. It can be changed by manipulating the knobs of different parameters provided in the calculator.

B. **Average {Activity} Footprint**

This output provides the average carbon footprint corresponding to the particular activity.

C. **Total Footprint**

This output presents the total carbon footprint based on all the activities included.

Here, \{Activity\} denotes one of the activities as mentioned earlier.

10.4.3 Carbon Footprint Calculator, United States Environmental Protection Agency

Carbon footprint calculator calculates the carbon emissions on three grounds, i.e., home energy, transportation, and waste [27]. It takes input in the form of various parameters for each of the three mentioned fields separately and simultaneously provides the following output:

A. **Your Current Total**

This output calculates the total carbon emissions for all the activities listed under home energy, transportation, and waste.

B. **New Total After Your Planned Actions**

This calculator provides recommendations regarding actions that need to be taken to minimize the carbon footprint. The output metric is the result of accomplishing the planned activities.

C. **US Average**

This output provides the average carbon emissions on the basis of basic information regarding family members and zip code, fed to the calculator.

10.4.4 *Footprint Calculator*

The footprint calculator, developed by Global Footprint Network to calculate the carbon emissions, is available in the form of a mobile application [28]. After taking an extensive survey using ratings, it provides a detailed report of ecological as well as individual carbon footprint. The ecological footprint is calculated using global hectares and the carbon footprint by tons per year.

10.5 Materials and Methods

This section discusses various aspects of the dataset, such as its collection, preprocessing, and application. It also provides information about the approaches used for visualizing and modeling the data.

10.5.1 *Dataset*

A dataset is created to propose a way to find suitable parameters that can bring down the overall carbon emissions score [29]. For this, the data is collected from the $MLCO_2$ calculator, which consists of four parameters, i.e., “hardware type”, “hours used”, “cloud provider”, and “region of compute” [8]. A comparison is made by fixing different parameters. “Hours used” is kept as constant since hours can be considered agnostic to the choice of optimal parameters. In order to have a proper visualization, eight hardware structures, out of the total 28 hardware, are considered for two different cases. In the first one, the “hours used” is kept to be 10 h and the “region of compute” is set as “US-Central1”, whereas “hardware” is considered as a variable. In the second one, “hours used” is fixed at 20 h, and the cloud provider is set as “Amazon Web services (AWS)”, and the “hardware” and the “region of compute” are treated as variables.

10.5.2 *Methods*

At first, a method is proposed to present the combination of parameters that emits the lowest carbon. For this method, three parameters, viz, “hardware”, “region of compute”, and “cloud provider” such as Google Cloud Platform, AWS, and Microsoft Azure are considered to visualize and compare. This method deals with plotting the information of these different parameters on the basis of their carbon emissions.

Next, the implemented real-time carbon emissions calculator for PC is used to conduct some experiments on calculating carbon emissions on a PC. The inspiration

Table 10.4 Python library and function for runtime

	CPU	GPU
Python library	psutil	GPUtil
Python library function	cpu_times	nvidia-smi

for the real-time carbon emissions calculator is derived from the "Green Algorithms" proposed by Loïc et al., which make use of system variables like power usage effectiveness, memory available, runtime, pragmatic scaling factor, etc. [29]. Influenced by the above research, an application is created to calculate the carbon emissions of a system in real time. The application constantly takes "runtime" and "memory" from the system in real time while keeping other parameters such as "cores", "Thermal Design Power (TDP)", "core type" as static. The parameters are briefly explained as follows:

Runtime: This is a dynamic parameter which provides information regarding the time taken by CPU or GPU to perform an action and is accountable for variation in carbon emissions. The runtime for CPUs and GPUs is fetched using the following Python library and function as shown in Table 10.4:

Memory: Available memory is the second dynamic parameter that also impacts carbon emissions from a PC. It may be noted here that the memory in use does not influence the carbon emissions. The Python library, which is used for getting the value of available memory for CPU and GPU, is the same as given in Table 10.4.

Cores: The carbon footprint of a PC is dependent on the number of cores which are responsible for improving the computational power. This is a static parameter and gets the value using the Python library mentioned in Table 10.4.

TDP: Carbon emissions are also the result of the amount of heat generated from a PC. Thermal Design Power (TDP) parameter is used to get the values of heat dissipated by the system. TDP values are directly proportional to the number of cores.

Core Type: Generally, there are two types of cores, CPU and GPU, responsible for the carbon emissions of a PC. This parameter is also considered as static.

10.6 Experiments and Results

This section presents the results of the experiments carried out. While the first sub-section shares the visualization results with various combinations of parameters, the second sub-section calculates the carbon emissions for a PC in real time.

10.6.1 Choosing the Right Parameters

In this section, an experiment has been carried out by feeding different sets of parameters to the $MLCO_2$ carbon emissions calculator to visualize the result. Figure 10.4 consists of a chart where carbon emissions are plotted by.

1. Keeping the "number of hours" as constant to "10 h".
2. Setting the "region of compute" as "US-Central1".
3. Varying "hardware provider" and "processing system" values.

The following figure basically gives an idea of the carbon footprint of various cloud providers under similar settings.

It is clear from Fig. 10.4 that Azure has the highest carbon footprint for similar processing systems, followed by Google Cloud and AWS. Hence, AWS can be considered as an efficient platform with the above set of parameters.

Figure 10.5 presents the carbon emissions of various processing systems in different regions of computing. Taking a clue from Fig. 10.4, the cloud provider is fixed to be AWS, and also the training time is set to 20 h.

It is clearly evident that the Asia–Pacific (Mumbai) region has the highest carbon emissions as compared to the other areas under study, i.e., AWS GovCloud (US), China (Beijing), and EU (Ireland). This points out to the direction that work corresponding to making the processing system more efficient is to be done primarily in the Asia–Pacific (Mumbai) region. The Asia–Pacific (Mumbai) region is followed by China (Beijing), EU (Ireland), and lastly AWS GovCloud (US) with regard to carbon emissions on similar processing systems.

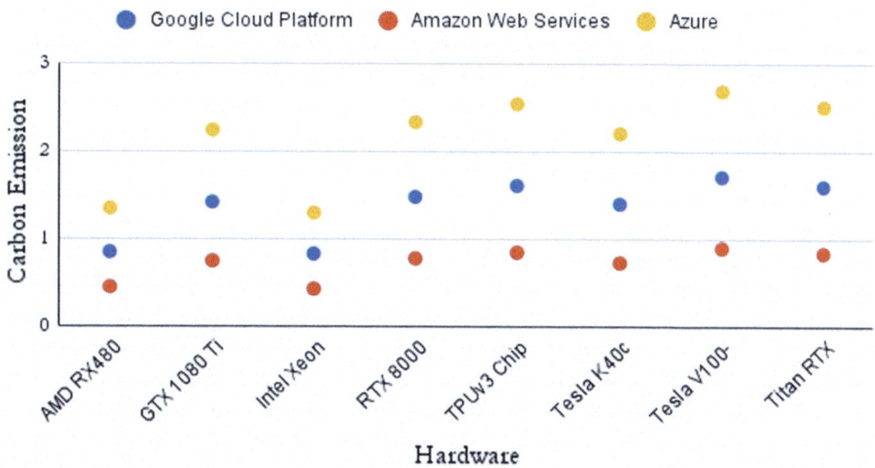

Fig. 10.4 Carbon emission comparison of different cloud providers on the basis of hardware

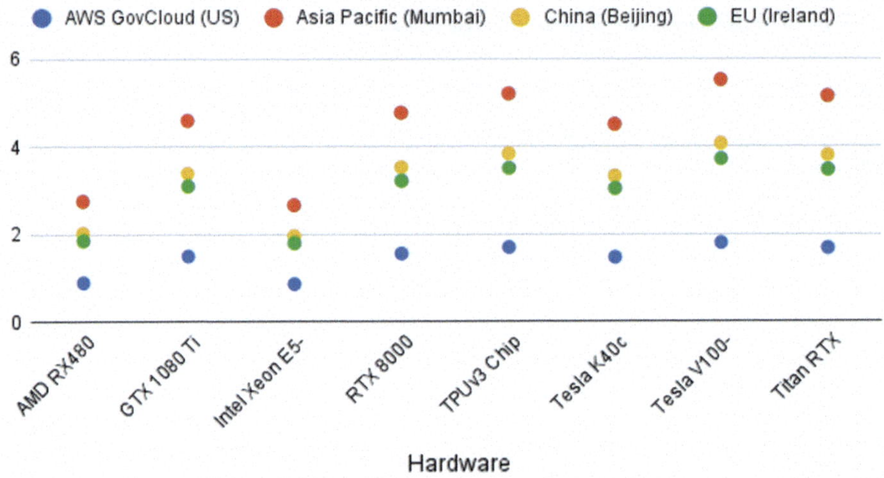

Fig. 10.5 Carbon emission comparison of AWS for different regions of compute on the basis of hardware

10.6.2 Real-Time Carbon Emissions Calculator

Since the configurations, i.e., runtime and memory, are dynamic in nature, the carbon emissions keep on changing. It would be great if the users get constant feedback about the carbon emissions from the PC while preparing an ML model or just casually working on it. Many researchers aspire to have a calculator that can inform them about the personal computer's carbon emissions with good precision so that they can use the system wisely. An attempt has been made to calculate the carbon emissions on a real-time basis from the personal computer. Table 10.5 discusses the carbon emission values both for the Central Processing Unit (CPU) and Graphical Processing Unit (GPU) on a different set of parameters. For the calculation of carbon emissions, Xeon E5-2670 (CPU) and TPU v3 (GPU) are considered.

10.7 Conclusions and Future Directions

In this study, we focus on elucidating the negative aspects of Artificial Intelligence towards climate change. Inference on a machine learning emissions calculator is also being made, and different combinations of hardware, region of computing, and cloud providers for machine learning models have been analyzed to get a glimpse of their carbon footprint. Moreover, the chapter also discusses a real-time carbon emissions calculator used to calculate the carbon emissions of personal computers.

The exploitation of nature has a major stake in the prosperity of human civilization. Ancient people cut trees to build houses, boats, etc. As human beings evolved,

Table 10.5 Carbon emission values for Xeon E5-2670 (CPU) and TPU v3 (GPU) on different sets of inputs

Runtime	Cores	Memory	TDP	Core type	Carbon emissions
27.7918	6	1.9803	14.4	CPU	1.2073 kg
27.7920	6	1.9812	14.4	CPU	1.2073 kg
27.7923	6	1.9797	14.4	CPU	1.2073 kg
27.7925	6	1.9853	14.4	CPU	1.2074 kg
27.7928	6	2.0006	14.4	CPU	1.2075 kg
20	1	0.2793	200	GPU	1.9952 kg
100.0012	1	0.3672	200	GPU	9.9778 kg
140.0018	1	1.1406	200	GPU	13.989 kg
180.0024	1	1.1368	200	GPU	17.9858 kg
200.0027	1	1.1368	200	GPU	19.9842 kg

they learnt various secrets about nature and exploited them to fulfil their needs. This process went on, and nature is exploited to the fullest in the modern world. However, it would have been difficult to thrive without using natural resources, but the endeavor should have been not only to avoid wastage but also to reproduce such resources. Besides, people are continuously polluting water bodies, getting indulged in reckless deforestation, and many more. Such actions lead to the deterioration of the environment and ultimately lead to an adverse climate.

Technology has also started to contribute predominantly towards the debilitation of climate. As more and more AI systems emerge with more accurate but not efficient solutions, the contribution will continue to burgeon. On reaching global warming levels of 1.5–2 °C above pre-industrial levels, it would not be easy for humankind to survive on the earth. Thus, the following standard processes are proposed to inculcate analytical activities in order to collectively undermine the stake of AI in climate change and attenuate the carbon footprint to achieve the net-zero target:

(a) A conscious effort should be taken to perform a smaller number of failed experiments in building ML models. The general practice is to run a model many times until the error is minimized and the model functions perfectly. However, the multiple iterations towards building a model for debugging could result in unnecessary carbon emissions. These trails could be minimized if models are prepared with fastidiousness and good quality data.
(b) An appropriate decision should be considered to deploy the model to a cloud provider after taking into account its efficiency of hardware and region of computing. As shown in the above experiment, different cloud providers have different carbon emissions for the same set of parameters. Therefore, choosing the right cloud provider for the right task thus becomes really crucial as it could prevent a lot of carbon emissions.
(c) Sending non-essential emails and messages through social media platforms should be avoided, as this simple mode of communication also involves

emissions of carbon. Other social media platforms also contribute to carbon emissions, and thus reckless use of such should be controlled.

(d) There is a need to strike a proper balance between improving the accuracy of the ML model and carbon emissions generated due to the same. As per the study mentioned above, an increase in 0.1 BLEU score costs USD 150,000.00. Hence, a trade-off between accuracy and cost should be considered.

Future work would deal with enhancing the current model to include real-time calculations for the cloud as well as local servers. Moreover, training, inference, and comparison of several ML models would be performed with Floating Point Operation (FPO) as the evaluation metric. Another work would include creating a dense dataset containing the TDP values of almost every CPU and GPU or finding a new way of calculating the TDP values dynamically for any system.

References

1. Fifth Generation' Became Japan's Lost Generation (1992) The New York Times, June 5, Section D, pp 1
2. Campbell, Murray A, Joseph Hoane Jr, Feng-hsiung Hsu (2002) Deep blue. Artific Intell 134.1-2:57–83
3. Fukushima K, Sei M (1982) Neocognitron: a self-organizing neural network model for a mechanism of visual pattern recognition. Competition and Cooperation in Neural Nets. Springer, Berlin, Heidelberg, pp 267–285
4. Dhar P (2020) The carbon impact of artificial intelligence. Nat Machine Intell 2:423–425
5. Lacoste A, Luccioni A, Schmidt V, Dandres T (2019) Quantifying the carbon emissions of machine learning. Available at http://arxiv.org/abs/1910.09700
6. Schwartz R, Dodge J, Smith NA, Etzioni O (2020) Green AI. Commun ACM 63:54–63
7. IPCC (2018) Summary for policymakers. In: Global Warming of 1.5 °C. An IPCC Special Report on the impacts of global warming of 1. 5°C above pre-industrial levels and related global greenhouse gas emission pathways, in the context of strengthening the global response to the threat of climate change, sustainable development, and efforts to eradicate poverty [Masson-Delmotte V, P Zhai, H-O Pörtner, D Roberts, J Skea, PR Shukla, A Pirani, W Moufouma-Okia, C Péan, R Pidcock, S Connors, JBR Matthews, Y Chen, X Zhou, MI Gomis, E Lonnoy, T Maycock, M Tignor, T Waterfield (eds.)]
8. https://github.com/mlco2/impact/tree/master/data
9. Strubell E, Ganesh A, McCallum A (2020) Energy and policy considerations for modern deep learning research. Proce AAAI Conf Artific Intell 34(09):3693–13696. https://doi.org/10.1609/aaai.v34i09.7123
10. Pidgeon N (2012) Public understanding of, and attitudes to, climate change: UK and international perspectives and policy. Climate Policy 12(sup01):S85–S106. https://doi.org/10.1080/14693062.2012.702982
11. Schmidt V, Luccioni A, Mukkavilli KS, Balasooriya N, Sankaran K, Chayes J, Bengio Y (2019) Visualizing the consequences of climate change using cycle-consistent adversarial networks. https://doi.org/10.48550/arXiv.1905.03709
12. Rolnick D, Donti PL, Kaack LH, Kochanski K, Lacoste A, Sankaran K, Ross AS, Milojevic-Dupont N, Jaques N, Waldman-Brown A, Luccioni A, Maharaj T, Sherwin ED, Mukkavilli KS, Kording KP, Gomes C, Ng AY, Hassabis D, Platt JC, Creutzig F, Chayes J, Bengio Y (2019) Tackling climate change with machine learning. https://doi.org/10.48550/arXiv.1906.05433

13. Patterson D, Gonzalez J, Le Q, Liang C, Munguia L, Rothchild D, So D, Texier M, Dean J (2021) Carbon emissions and large neural network training. https://doi.org/10.48550/arXiv.2104.10350
14. Durgam DK, Sao S, Singh RK (2017) Effect of mobile tower radiation on birds in Bijapur district, Chhattisgarh. World J Pharm Pharmac Sci 6:1221–1229
15. Amazon Sustainability 2020 Report: Further and Faster Together. https://sustainability.aboutamazon.com/pdfBuilderDownload?name=amazon-sustainability-2020-report
16. Bergstra J, Bengio Y (2012) Random search for hyper-parameter optimization. J Mach Learn Res 13:281–305
17. So D, Le Q, Liang C (2019) The evolved transformer. International Conference on Machine Learning. Proceedings of the 36th International Conference on Machine Learning, PMLR, 97: pp 5877–5886
18. Koten H, Bilal S (2018) Recent developments in electric vehicles. Intern J Adv Autom Technol 1(1):35–52
19. André Gonçalves. Are Electric Cars Really Greeners. https://youmatter.world/en/are-electric-cars-eco-friendly-and-zero-emission-vehicles-26440/
20. Ahmed M, Zheng Y, Amine A, Fathiannasab H, Chen Z (2021) The role of artificial intelligence in the mass adoption of electric vehicles. Joule 5(9):2296–2322. https://doi.org/10.1016/j.joule.2021.07.012
21. Curran C (2020) What will 5G mean for the environment? https://jsis.washington.edu/news/what-will-5g-mean-for-the-environment/
22. Amy N, Kristen C (2017) Everything you need to know about 5G: millimeter waves, massive MIMO, full duplex, beamforming, and small cells are just a few of the technologies that could enable ultrafast 5G networks. IEEE Spectrum. January 27
23. The thought experiment: What is the carbon footprint of an email?, Science Focus (2020). https://www.sciencefocus.com/planet-earth/the-thought-experiment-what-is-the-carbon-footprint-of-an-email/
24. Cook G, Lee J, Tsai T, Kongn A, Deans J, Johnson B, Jardim B (2017) Clicking clean: who is winning the race to build a green internet? Technical report, Greenpeace
25. Center for Sustainable Systems, University of Michigan (2021) Carbon Footprint Factsheet. Pub. No. CSS09–05
26. https://www.nature.org/en-us/get-involved/how-to-help/carbon-footprint-calculator/
27. https://www3.epa.gov/carbon-footprint-calculator
28. https://www.footprintnetwork.org/resources/footprint-calculator/
29. Lannelongue L, Grealey J, Inouye M (2021) Green algorithms: quantifying the carbon footprint of computation. Adv Sci. https://doi.org/10.1002/advs.202100707

The manufacturer's authorised representative in the EU is Springer Nature Customer Service Centre GmbH, Europaplatz 3, 69115 Heidelberg, Germany. If you have any concerns regarding our products, please contact ProductSafety@springernature.com

Printed and bound by CPI Group (UK) Ltd, Croydon, CR0 4YY
30/09/2026
02225315-0005